Research Reports in Physics

Research Reports in Physics

Nuclear Structure of the Zirconium Region
Editors: J. Eberth, R. A. Meyer, and K. Sistemich

Ecodynamics Contributions to Theoretical Ecology
Editors: W. Wolff, C.-J. Soeder, and F. R. Drepper

Nonlinear Waves 1 Dynamics and Evolution
Editors: A. V. Gaponov-Grekhov, M. I. Rabinovich, and J. Engelbrecht

Nonlinear Waves 2 Dynamics and Evolution
Editors: A. V. Gaponov-Grekhov, M. I. Rabinovich, and J. Engelbrecht

Nonlinear Waves 3 Physics and Astrophysics
Editors: A. V. Gaponov-Grekhov, M. I. Rabinovich, and J. Engelbrecht

Nuclear Astrophysics
Editors: M. Lozano, M. I. Gallardo, and J. M. Arias

Optimized LCAO Method and the Electronic Structure of Extended Systems
By H. Eschrig

Nonlinear Waves in Active Media   Editor: J. Engelbrecht

Problems of Modern Quantum Field Theory
Editors: A. A. Belavin, A. U. Klimyk, and A. B. Zamolodchikov

Fluctuational Superconductivity of Magnetic Systems
By M. A. Savchenko and A. V. Stefanovich

Nonlinear Evolution Equations and Dynamical Systems
Editors: S. Carillo and O. Ragnisco

Nonlinear Physics   Editors: Gu Chaohao, Li Yishen, and Tu Guizhang

Nonlinear Waves in Waveguides with Stratification   By S. B. Leble

Quark-Gluon Plasma   Editors: B. Sinha, S. Pal, and S. Raha

Symmetries and Singularity Structures
Integrability and Chaos in Nonlinear Dynamical Systems
Editors: M. Lakshmanan and M. Daniel

Modeling Air-Lake Interaction Physical Background
Editor: S. S. Zilitinkevich

Nonlinear Evolution Equations and Dynamical Systems NEEDS '90
Editors: V. G. Makhankov and O. K. Pashaev

Solitons and Chaos
Editors: I. Antoniou and J. F. Lambert

Electron-Electron Correlation Effects in Low-Dimensional Conductors and Superconductors Editors: A. A. Ovchinnikov and I. I. Ukrainskii

A. A. Ovchinnikov   I. I. Ukrainskii (Eds.)

Electron-Electron
Correlation Effects in
Low-Dimensional Conductors and Superconductors

With 41 Figures

Springer-Verlag
Berlin Heidelberg New York London Paris
Tokyo Hong Kong Barcelona Budapest

Professor Dr. Alexandr. A. Ovchinnikov
Chemical Physics Institute, Academy of Science of the USSR, Kosygin Street 4,
SU-117334 Moscow B-334, USSR

Professor Dr. Ivan I. Ukrainskii
Institute for Theoretical Physics, Metrologicheskaya 14-B,
SU-252143 Kiev 143, USSR

ISBN 3-540-54248-5 Springer-Verlag Berlin Heidelberg New York
ISBN 0-387-54248-5 Springer-Verlag New York Berlin Heidelberg

Library of Congress Cataloging-in-Publication Data. Electron-electron correlation effects in low-dimensional conductors and semiconductors / A. A. Ovchinnikov, I. I. Ukrainskii, eds. p. cm. – (Research reports in physics) Proceedings of a meeting held at the Institute for Theoretical Physics in Kiev May 15–18, 1990. Includes bibliographical references and indexes. ISBN 0-387-54248-5 1. One-dimensional conductors–Congresses. 2. Superconductors–Chemistry–Congresses. I. Ovchinnikov, A. A. (Aleksandr Anatol'evich) II. Ukrainskii, I. I. (Ivan I.), 1943– . III. Instytut teoretychnoi fizyky (Akademiia nauk Ukrains'koi RSR) IV. Series. QC176.8.E4E35 1991 537.6'2–dc20 91-30967

This work is subject to copyright. All rights are reserved, whether the whole or part of the material is concerned, specifically the rights of translation, reprinting, reuse of illustrations, recitation, broadcasting, reproduction on microfilm or in any other way, and storage in data banks. Duplication of this publication or parts thereof is permitted only under the provisions of the German Copyright Law of September 9, 1965, in its current version, and permission for use must always be obtained from Springer-Verlag. Violations are liable for prosecution under the German Copyright Law.

© Springer-Verlag Berlin Heidelberg 1991
Printed in Germany

The use of general descriptive names, registered names, trademarks, etc. in this publication does not imply, even in the absence of a specific statement, that such names are exempt from the relevant protective laws and regulations and therefore free for general use.

Typesetting: Data conversion by Springer-Verlag
57/3140-543210 – Printed on acid-free paper

Preface

Advances in the physics and chemistry of low–dimensional systems have been really magnificent in the last few decades. Hundreds of quasi–one–dimensional and quasi–two–dimensional systems have been synthesized and studied.

The most popular representatives of quasi–one–dimensional materials are polyacethylenes CH_x [1] and conducting donor–acceptor molecular crystals TTF–TCNQ. Examples of quasi–two–dimensional systems are high temperature superconductors (HTSC) based on copper oxides LA_2CuO_4, $YBa_2Cu_3O_{6+y}$ and organic superconductors based on BEDT–TTF molecules.

The properties of such one– and two–dimensional materials are not yet fully understood. On the one hand, the equations of motion of one–dimensional systems are rather simple, which facilitates rigorous solutions of model problems. On the other hand, manifestations of various interactions in one–dimensional systems are rather peculiar. This refers, in particular, to electron–electron and electron–phonon interactions. Even within the limit of a weak coupling constant electron–electron correlations produce an energy gap in the spectrum of one–dimensional metals implying a Mott transition from metal to semiconductor state.

In all these cases perturbation theory is inapplicable. Which is one of the main difficulties on the way towards a comprehensive theory of quasi–one–dimensional systems. – This meeting held at the Institute for Theoretical Physics in Kiev May 15-18 1990 was devoted to related problems. The papers selected for this volume are grouped into three sections.

Section 1 contains papers dealing with electron correlation problems. Applications to 2–d superconductors, a connection between 1–d Hubbard and Luttinger models, a treatment of correlations in high–T_c superconductors, the 2–d Peierls–Hubbard model giving rise to kink–antikink pairs with superconducting condensation and the optical spectra of superconducting copper–oxides are discussed. *Section 2* provides information on progress in the understanding of Mott-Peierls semiconductor polymers, on the way in which electron–electron correlations affect the properties of polymers and a description of a new approach to the study of electron–electron correlations in large molecules and polymers. *Section 3* deals with correlation effects in magnetization and kinetics including some features of the 2–d Hubbard model.

Kiev,
December 1990

A.A. Ovchinnikov
I.I. Ukrainskii

Contents

Introduction
By A.A. Ovchinnikov and I.I. Ukrainskii 1

| Part I | Correlation Effects in Low-Dimensional Conductors, Superconductors and Model Systems |

The 1–d Hubbard Model: A Landau Luttinger Liquid
By J. Carmelo and A.A. Ovchinnikov 12

Mean-Field Study of Possible Electronic Pairings
in the CuO Plane of HTSO
By A.A. Ovchinnikov and M.Ya. Ovchinnikova
(With 2 Figures) 23

Correlation Pairing and Antiferromagnetic Phase Energy
in Low–Dimensional Systems of La–Sr–Cu–O and Y–Ba–Cu–O
Metaloxides
By I.I. Ukrainskii and E.A. Ponezha (With 8 Figures) 32

Kink Nature of Current Carriers
in High–T_c Superconductor Oxides
By I.I. Ukrainskii, M.K. Sheinkman, and K.I. Pokhodnia
(With 3 Figures) 41

Anomaly Index and Induced Charge on a Noncompact Surface
in an External Magnetic Field
By Yu.A. Sitenko 48

About the Influence of Uniaxial Pressure on the Twin Structure
in the 1–2–3 System
By V.S. Nikolayev (With 4 Figures) 54

Part II	Correlation Effects in Organic Crystals, Molecules and Polymers

Coexistence of Mott and Peierls Instabilities
in Quasi–One–Dimensional Organic Conductors
By I.I. Ukrainskii and O.V. Shramko (With 5 Figures) 62

Nonlinear Optical Susceptibility for Third Harmonic Generation
in Combined Peierls Dielectrics
By Yu.I. Dakhnovskii and K.A. Pronin (With 1 Figure) 73

Nonlinear Optical Properties of $(A - B)_x$–Polymers
By Yu.I. Dakhnovskii and A.D. Bandrauk (With 1 Figure) 80

Application of the Method of Cyclic Permutations
to the Calculation of Many–Electron Systems.
Polaron States in the Emery Model
By V.Ya. Krivnov, A.A. Ovchinnikov, and V.O. Cheranovskii
(With 3 Figures) 86

From Incomplete Allowance for Electron Correlation
to the Full CI in π–Systems. The Variational Operator Approach
By A.V. Luzanov, Yu.F. Peash, and V.V. Ivanov 93

Dynamical Correlation in Finite Polymethine Chains
By G.G. Dyadyusha and I.V. Repyakh (With 1 Figure) 100

Electronic Structure and Optical Spectra of Transition Metal
Complexes via the Effective Hamiltonian Method
By A.V. Soudackov, A.L. Tchougreeff, and I.A. Misurkin ... 106

Part III	Multiparticle Effects in Kinetics and Magnetism

Magnetic Properties of the Hubbard Model
with Infinite Interactions
By V.Ya. Krivnov, A.A. Ovchinnikov, and V.O. Cheranovskii
(With 4 Figures) 114

Anomalous Transport Through Thin Disordered Layers
By S.F. Burlatsky, G.S. Oshanin, and A.I. Chernoutsan
(With 4 Figures) 121

Correlation Effects in Many–Body Reactive Systems
By S.F. Burlatsky, G.S. Oshanin, and A.A. Ovchinnikov 129

Fermionization of a Generalized Two-Dimensional Ising Model
By A.I. Bugrij (With 5 Figures) 135

Ferromagnetism of Charge–Transfer Crystals:
Curie Temperature of a Organometallic Ferromagnet
By A.L. Tchougreeff and I.A. Misurkin 152

Subject Index 159

Index of Contributors 161

Introduction

A.A. Ovchinnikov[1] and I.I. Ukrainskii[2]

[1] Institute of Chemical Physics, Kosygin St. 4, Moscow 117334, USSR
[2] Institute for Theoretical Physics, Metrologicheskaya 14, SU-252130 Kiew, USSR

Advances in physics and chemistry of low–dimensional systems have been really magnificent in the last few decades. Hundreds of quasi–one–dimensional and quasi–two–dimensional systems have been synthesized and studied. The Properties of those materials attracted physicists, chemists and engineers.

The most popular representatives of quasi–one–dimensional materials are polyacetylenes CH_x [1] and conducting donor–acceptor molecular crystals TTF–TCNQ [2].

One of the promising families relates to quasi–two–dimensional systems are new high temperature superconductors (HTSC) based on copper oxides LA_2CuO_4, $YBa_2Cu_3O_{6+y}$ [3] and organic superconductors based on BEDT–TTF molecules [4].

Quantum processes in low–dimensional systems are characterized by a number of peculiarities. This fact results in the development of special methods of theoretical studies in low–dimensional phenomena. We describe this problem now for one–dimensional (1–d) systems. In one–dimensional physics and chemistry there is a number of certain difficulties and some of them are far from being overcome. On the one hand, motion equations in one–dimensional systems are much simpler. This facilitates rigorous solution of a model problem which is often impeded by a large number of dimensions. On the other hand manifestations of various interactions in one–dimensional systems are rather peculiar. This refers, in particular, to electron–electron and electron–phonon interactions. The perturbation theory is inapplicable in both cases. Thus, electron–phonon interaction leads to field localization of electron excitation in one–dimensional system which results in soliton excitations and Peierls deformation. Calculations of soliton excitation can not be done by decomposition in the series of electron–phonon coupling constant.

Electron–electron interactions, even within the limit of a weak coupling constant, produces an energy gap in the spectrum of one–dimensional metal which means the Mott transition from metal to semiconductor state. And in this case the perturbation theory is inapplicable.

Similar situation occurs in one–dimension with respect to electron–impurity interactions. Started by Mott and Twose theoretical studies of this problem show that all one–electron states in 1–d disordered system are localized and, hence, cannot be calculated using the perturbation theory. State localization turns the direct current conductivity into zero.

Inapplicability of the perturbation theory is one of the main difficulties on the way to accomplish the theory of quasi–one–dimensional systems.

These difficulties were being surpassed in different ways.

Regarding electron–phonon interaction the most fruitful method is is to reduce the set of equations into a completely integrable system which can be the nonlinear Schrödinger equation, the sine–Gordon equation and others.

Advances in description with respect to electron–electron interactions turned out to be less pronounced however more yielding regarding the physics of 1–d systems. The major reason for it lies in well–known complications of the many–electron theory for systems with an infinitely number of electrons.

Quantum chemistry and, in particular, the theory of many–electron systems are based upon the Hartree–Fock approximation. Making a joke theoreticians often rephrase the saying "the word came first" into "the Hartree–Fock approximation came first".Then various many–electron theories appeared where the wave function must be represented not by one Slater determinant but an infinite series of these determinants. And if the number of particles in the system grows as $N(N \to \infty)$ then the number of terms in this infinite series must increase at least as e^{an}, where a is a constant ($a \approx 1$). This particular infinite complication of the theory is the main hindrance for its wide application in calculations. One of the objectives of the present book is to show , however that very often these difficulties are being considerably exaggerated. As a rule, having analyzed the Hamiltonian of the system under study using the many–electron theory one can reduce the problem to a simpler Hamiltonian or, or without any loss in quality construct multiconfigurational wave function of the system which can be factorized into an antisymmetrized product of one– or two–electron wave functions. As approximations for a wave function, besides the extended Hartree–Fock approximation (EHF) described in details in [1], the spinless fermion approximation in case of strong interactions and the variable localized geminals approximation (VLG) can be mentioned [5].

In the EHF and spinless fermion approximations a many–electron wave function is finally factorized into the product of single–particle functions (orbitals), and in the VLG approximation the factorization into the product of two–particle functions (geminals) is done.

Now we draw the reader's attention to another aspect of the theory of quasi–one–dimensional systems. Real systems with one–dimensional anisotropy are, in fact, three–dimensional. In case of a theoretical study it is expedient to mentally separate a 1–d subsystem out of the real system using its specific features.

The separation of a quasi–one–dimensional subsystem goes naturally through analysis of the total Hamiltonian represented by the sum

$$\widehat{\mathcal{H}} = \sum_n \widehat{\mathcal{H}}_n + \frac{1}{2} \sum_{n,m} \widehat{V}_{n,m} , \qquad (1)$$

where $\widehat{\mathcal{H}}_{n,m}$ is the Hamiltonian of a n-th quasi–one–dimensional substructure (filaments needles, chains or stacks), and operators $\widehat{V}_{n,m}$ describe its interactions with other quasi–one–dimensional subsystems.

Further we assume that the interaction operators include no terms responsible for electron exchange between separate quasi–one–dimensional subsystems. And this predetermines the subdivision of the Hamiltonian into the sum (1). This approximation provides satisfactory description of polyacetylenes, donor–acceptor molecular conducting crystals and many other quasi–one–dimensional systems.

Before we start to consider particular expressions for the Hamiltonians of electron–phonon systems under study it is worthwhile to note the following.

Most processes of interest in quasi–one–dimensional systems are determined by the energy spectrum and the nature of elementary excitations of these systems. The low–energy region of the spectrum is mainly related to a small part of the total number of the system electrons. This facilitates a rigorous enough description of electron processes occurring in these systems. Say, most interesting properties of polyacetylenes originate from the π–electrons number equals the number of carbon atoms and essentially less than the total number of electrons in the system.

Studying most significant properties of donor–acceptor molecular conducting crystals it suffices to consider one electron only per a donor–acceptor pair. For a TTF–TCNQ crystal it means that only one electron out of 208 is to be considered.

A similar situation occurs in other quasi–one–dimensional systems.

A reduced number of the degrees of freedom of the system requires a model Hamiltonian parameters of such a Hamiltonian are to be obtained from experiments on related systems.

In terms of quantum chemistry we should use semi–empirical methods.

Coming back to close similarity in the quantum chemistry and the theory of conjugated molecules and polymers, we write down a well known Hückel–Pople (HP) Hamiltonian [1]

$$\mathcal{H} = \sum_{mm'\sigma} \beta_{mm'} c^+_{m\sigma} c_{m'\sigma} + \sum_m \alpha_m c^+_{m\sigma} c_{m\sigma} + \frac{1}{2} \sum_{mm'\sigma} \gamma_{mm'} c^+_{m\sigma} c_{m\sigma} c^+_{m'\sigma} c_{m'\sigma} , \qquad (2)$$

where $c_{m\sigma}^+$ is an electron creation operator on the m-th site with the spin σ. The HP Hamiltonian allowed to study most interesting properties of molecules with conjugated bonds.

As a rule, in (2) it suffices to consider electron hopping (a resonance term) for adjacent atoms only, that is

$$\beta_{mm'} = \beta(R_{m,m'})\delta_{m',m+1}, \qquad (3)$$

where $R_{m,m'}$ is the distance between m-th and m'-th atoms. To obtain a qualitatively correct description of most effects one can use in the electron interaction operator (the second term in (2)) only first several terms. A most frequently used approximation is that of Hubbard–Anderson (HA)

$$\gamma_{mn} = \gamma_0 \delta_{mn}. \qquad (4)$$

Sometimes it is of importance to consider electron interactions of the neighboring atoms, that is, to assume

$$\gamma_{mn} = \begin{cases} \gamma_0 \delta_{mn}, \\ \gamma_1 \delta_{m\pm 1,n}. \end{cases} \qquad (5)$$

The values of a resonance integral (3) is a function of the distance between the m-th and $(m+1)$-th sites. In conjugated molecules it is sufficient to use the first term of this function expansion in the vicinity of $R = R_0 = 1.397$ Å corresponding to the c–c bond length in benzene

$$\beta(R) = \beta_0 - (R - R_0)\beta'. \qquad (6)$$

The approximations (3)–(6) suffice to study a great number of experiments except for dynamic (and kinetic) properties of quantum–one–dimensional systems to explicitly account for vibrational degrees of freedom. This is done by adding to the Hamiltonian (2) the phonon Hamiltonian

$$\mathcal{H}_{ph} = \sum_{ki} \hbar \Omega_{ki} (b_{ki}^+ b_{ki} + \frac{1}{2}), \qquad (7)$$

where b_{ki}^+ is a phonon creation operator of the i-th mode with a quasi-momentum k.

Starting from (6), the operator of electron–phonon subsystem interactions is chosen like that suggested by Fröhlich

$$\mathcal{H}_{e-ph} = \sum_{kq} \lambda_{qi} (b_{q,i}^+ - b_{-q,i}) a_{k,\sigma}^+ a_{k+q,\sigma}, \qquad (8)$$

where a constant λ_{ki} is proportional to the β derivative with respect to R, that is β' in (6).

Like other cases, for quasi–one–dimensional systems it is often sufficient to use only the classical form of the phonon part of the Hamiltonian

$$\mathcal{H}_{\mathrm{ph}} = \frac{1}{2} \sum_{mi} M_i \dot{\boldsymbol{R}}_{mi}^2 + \frac{1}{2} \sum_{mi} K_i (\boldsymbol{R}_{mi} - \boldsymbol{R}_{m+1,i})^2 \,, \qquad (9)$$

and

$$\mathcal{H}_{\mathrm{e-ph}} = \sum_{m\sigma} (R_0 - R) \beta' (c_{m+1,\sigma}^+ c_{m\sigma} + \mathrm{h.c.}) \,. \qquad (10)$$

The Hamiltonian (2) together with the expressions for matrix elements (3–5) allows us to consider the properties of materials based on conjugated polymer and of donor–acceptor molecular crystals with quasi–one–dimensional conductivity such as the crystals based on TCNQ–TTF and their derivatives TSF, TST and HTSC [1–4].

We can state that using the Hamiltonians mentioned in (1)–(10) we can explain all well known by now properties of a number of low–dimensional systems, considered in this book.

Taking in mind papers on electron–electron correlations in the Hubbard model (5) we now briefly mention the classification of electron terms of quasi–one– and quasi–two–dimensional systems. This classification will be needed further on. Besides conventional classification by multiplicity we shall distinguish between quasi–ionic and quasi–homeopolar levels can be traced back to Bogolubov's works. Let us consider a chain of N sites (atoms). Each of the site contains one electron. In the limit $\beta_m \to 0$ all the states can be subdivided with respect to the number of ionized sites [1].

The first group will contain the states with one electron per note, that is, only neutral sites. These states are homeopolar and their number is 2^N.

The second group refers to the states with one ionized pair of nodes whose number is $2^{N-2}(N-1)N$. Then, ordered with respect to energy, follows the group $(2!)^{-2} 2^{N-4} N(N-1)(N-2)(N-3)$ with two ionized pairs etc.

According to the Hubbard–Anderson approximation at $\beta_m = 0$ the states inside of each group are degenerated. Their energy is $p\gamma$, where p is the number of ionized node pairs.

At $\beta_m \ll \gamma$ the splitting of levels in the first group can be described by the Heisenberg Hamiltonian [1]. To calculate the energy of other levels the long–range part of electron interactions is of significance, that is, the terms with γ_{mn} in (2) ($m \neq n$).

Using real parameter values $4\beta_m \approx \gamma_{ma}$ energy levels can be also referred to one of these groups. And then the prefix "quasi" is added to the group name.

The assignment to the group of quasihomeopolar states means that within a dissociation limit these levels will belong to the group of homeopolar levels.

Energy distribution of level groups depends considerably on the Hamiltonian (2) parameters. In case of non–interacting electrons all the level groups are mixed up.

According to the Hubbard approximation the lowest levels are homeopolar ones, are quasi–ionic (current) excitations. The introduction of neighbor–site electron repulsion separate the domain of quasi–ionic currentless excitations (optically active excitations) from that of current excitations.

The fact, that excitations exist in polydiacetilenes and polyacetilenes near the threshold of current state band is another proof of the significance of correlation effects in polymers with conjugated bonds.

The greatest interest with respect to newly synthesized quasi–one–dimensio- nal and quasi–two–dimensional systems is attached to the compounds with a high electric conductivity.

But on the way to create good organic conductors the investigators encounter difficulties of not only technical but principal nature which relates to an electron instability of a conducting state.

To overcome these difficulties we must know the specify of an electron structure of quasi–one–dimensional crystals. Their most important peculiarity lies in the fact that a metallic state of a quasi–one–dimensional crystal is unstable with respect to a transition (at a temperature decrease) into a dielectric or semiconducting phase. The character of instability and its force strength which determines the metal–insulator transition temperature depends on structural features of the crystal.

Let us consider a system consisting of long needle packed into a 3–d crystal. The Hamiltonian of each needle is supposed to be the first term in the general expression (1)

$$\mathcal{H} = -\beta_0 \sum_{m\sigma=1}^{N} (c_{m\sigma}^+ c_{m+1,\sigma} + \text{h.c.}) , \qquad (11)$$

where the notifications are the same as in (2) and $N \to \infty$. A 1–d system with Hamiltonian (11) is a metal independently on the number of electrons in the conduction band N_e or on their density

$$\rho = \frac{N_e}{N} = \frac{1}{N} \sum_{m\sigma} \langle c_{m\sigma}^+ c_{m\sigma} \rangle , \qquad (12)$$

that is, at any filling of the conduction band $0 < \rho < 2$.

In case, when the number of electrons and sites coincide we have a half–filled conduction band, $N_e = N$ and the Fermi momentum is $k_F = \pi/2a$ where a is a 1–d lattice parameter.

A 1–d metal with a half–filled conduction band is unstable with respect to the following metal insulator transitions:

1) The Mott metal–insulator transition resulting from electron interactions. Instability of a 1–d metal with respect to this transition arises from the fact that electron–electron interactions produce the gap at $T = 0°$ K even within a weak coupling constant $U = \gamma/\beta_0$ in the Hamiltonian (2).

2) The Peierls metal–insulator transition referred to electron–phonon interactions. Alongside with the gap a periodic deformation of the crystal occurs with the period π/k_F.
3) The Anderson metal–insulator transition resulting from structure disordering of the crystal. The instability of a 1–d metal in this case is stipulated by localization of electron states even by a weak random field.

To the same class we attribute Wigner ordering of electrons in quasi–one–dimensional conductor which appears at large coupling parameter U [6].

Early theories of quasi–one–dimensional systems led to a conclusion [1] that various instabilities in a 1–d–metal are competing.

However, further analysis showed that, in fact, a coexistence of instabilities is possible.

Thus, in [7] it was shown that Mott and Peierls instabilities coexist both at $\rho = 1$ and at $\rho = 1/2$. In other words, a 1–d Mott insulator also undergoes lattice deformation with the period $2\pi/2k_F$.

If we want to obtain a good organic conductor or even superconductor we should stabilize the system with respect to the above transitions.

The history of quasi–one–dimensional metal synthesis is, in fact, the history of fighting the above instabilities.

One of the effective means to fight the metal–insulator transitions is to shift electron density ρ from the values approaching $1, 1/2, 1/3$ and other fractions with small denominators. This is done by crystal doping with electron donors or acceptors or by violation of a simple steahiometric ratio $\rho = 1$.

To understand why this simple and clear method is so efficient we shall discuss the instabilities and their descriptions more in detail.

We shall consider a system with a half–filled band $\rho = 1$.

The Mott metal–insulator transition. A system with Hamiltonian (11) at $\rho = 1$ a metal. Adding to (11) an interaction operator like (1) we obtain the system with Hubbard Hamiltonian

$$\mathcal{H}_x = \sum_{m\sigma}\{(-\beta)[c^+_{m\sigma}c_{m+1,\sigma} + \text{h.c.}] \\ + \frac{1}{2}\beta_0 U c^+_{m\sigma}c_{m\sigma}c^+_{m-\sigma}c_{m-\sigma} \ . \tag{13}$$

The spectrum of a cyclic chain with the Hamiltonian (13) is the spectrum of an insulator at any $U > 0$, that is, an excitation of states with charge transfer requires an energy ΔE_i.

For the first time a conclusion on the energy gap formation in such a system appeared in calculation by the extended Hartree–Fock (EHF) calculations were applied ([1] and references therein).

Further this conclusion was supported by exact calculations using the Bethe ansatz [1].

Calculations of an infinite chain using the Bethe Anzatz and EHF are described in detail in [1]. Further development of exact calculations for the Hamiltonian (3) also is presented in [1]. Quite a number of publications appeared on the EHP application to calculations of the Mott transition [1].

Peierls transition. Let us consider a system with the Hamiltonian which can be represented as the sum of (9), (10) and (11) at $R_m = 0$

$$\mathcal{H} = \sum_{m\sigma}[-[\beta_0 + \beta'(\boldsymbol{R}_m - \boldsymbol{R}_{m+1})]c^+_{m\sigma}c_{m+1,\sigma} + \text{h.c.}] + \frac{1}{2}k\sum_m (\boldsymbol{R}_m - \boldsymbol{R}_{m+1})^2. \tag{14}$$

The energy minimum of an infinite chain is reached by the Hamiltonian (14) when

$$R_m = R_0 \cos(Qam + \varphi_0), \tag{15}$$

where a is a non-deformed lattice parameter, x_0 is an amplitude of periodic deformation, φ_0 is the phase of this deformation, and $Q = 2k_F$, $\hbar k_F$ is the Fermi momentum.

For a half-filled band $k_F = \pi/2a$ and

$$2R_0 = 4\frac{\beta_0}{\beta'}\exp\left[-\frac{\pi\beta k}{(\beta')^2}\right]. \tag{16}$$

The energy spectrum of conduction electrons for a half-filled band is described by the expression

$$\varepsilon_{1,2} = \pm 2\beta_0 \sqrt{\cos^2 k + 4\left(\frac{\beta'}{\beta_0}\right) R_0^2 \sin^2 k}, \tag{17}$$

where "-" corresponds to a completely filled conduction sub-band, and "+" corresponds to a vacant sub-band.

This, the gap in the one-particle spectrum is

$$\Delta E_g = 8\beta' R_0. \tag{18}$$

The ground state energy correction is

$$\Delta E_c = 4\beta R_0^2 \ln R_0 + \frac{1}{2}K R_0^2, \tag{19}$$

where R_0 is preset by (14).

The shape of atom shift is, strictly speaking, a hypothesis. Recently it become clear that the problem like (14) can be solved exactly. Some specific features of physics in one-dimension remain valid also in two dimensions. But theoretical treatment of 2-d models is more complicated. Thus, the Mott and Anderson metal-insulator transitions can occur also in quasi-two-dimensional systems. However, the Peierls transition in 2-d case can appear only for special forms of the Fermi surface in the case of sv called nesting. Generally speaking, the conditions for the metal–to–insulator transitions in 2–d systems are stronger than those in 1–d case.

So, passing to 2–d systems we can stabilize conducting and superconducting states. The papers in this book presumably read with the systems described by Hamiltonians in the form (1) and (2). So, we can see how the theory works.

The papers are subdivided between three sections.

Section 1 consists of paper dealing with general electron correlation problems with applications to 2–d superconductors. An interesting coupling between 1–d Hubbard and Luttinger models we can find in paper by A.O. Ovchinnikov and J. Carmelo. The treatment of correlation effects in high-T_c superconductors on the basis of Hamiltonian (1),(2) we can see in paper by A. Ovchinnikov and M. Ovchinnikova. The Peierls–Hubbard model in 2–d case leads to kink–antikink pairs with superconducting condensation. I. Ukrainskii et al. use these circumstances to describe the optical spectra of superconducting copper–oxides.

Some properties of polymers which are the Mott–Peierls semiconductors consider in Section 2 in papers by I. Ukrainskii, Yu. Dakhnovskii et al. It follows from these papers that electron–electron correlations strongly affect the properties of polymers. This section contain also description of some new approach for study of electron–electron correlations in large molecules and polymers.

Section 3 deals with correlation effects in magnetization and kinetics. Some features of the 2–d Hubbard model are studied in the papers by V. Krivnov, A. Ovchinnikov.

References

1. A.A.Ovchinnikov, I.I.Ukrainskii, G.F.Kvenstsel. Soviet Physics Uspekhi, **15** 575 (1973)
2. Physics in One Dimension. J. Bernasconi, T. Schneider, Eds. (Springer, Berlin–Heidelberg–New-York 1981)
3. J.C. Bednorz, K.A. Müller. Z. Phys. **B64** 189 (1986)
4. E.B. Yagubskii, I.F. Schegolev et al. ZHETP Pisma **39** 12 (1984)
5. I.I. Ukrainskii. Theor. Math. Phys. **32** 392 (1977)
6. V.E. Klymenko, V.Ya. Krivnov, A.A. Ovchinnikov, I.I. Ukrainskii. J. Phys. Chem. Solids **39** 359 (1978)
7. I.I. Ukrainskii ZHETP **76** 760 (1979)

Part I

Correlation Effects in Low-Dimensional Conductors, Superconductors and Model Systems

The 1–d Hubbard Model: A Landau Luttinger Liquid

*J. Carmelo[2], A.A. Ovchinnikov[3]

[2] Max–Planck–Institut für Festkörperforschung, 7000 Stuttgart 80, FRG
[3] Institute of Chemical Physics, Kosygin St. 4, Moscow 117 334, USSR

1. Introduction

The full understanding of the Lieb and Wu solution for the 1–d Hubbard model [3] is of interest in its own right, and may provide clues to the understanding of higher dimensional systems [4]. In fact, the 1–d Hubbard Hamiltonian is a cornerstone of many–body theory, its exact solution providing a crucial test for approximate theories. The model belongs to a class of many–particle system which Haldane [5] has called Luttinger liquids, in which the spectral properties are regulated by unusual non–classical exponents. Anderson's claim that the $2 - D$ Hubbard model maps onto the 1–d one in each separate angular momentum channel [4], has also contributed for the present renewed interest on the model.

In this paper we use a new representation of the Lieb and We solution to study the properties of the 1–d Hubbard model in terms of the interaction of charge and spin pseudoparticles (p-particles). In addition to a new understanding of the spectral properties of the model [1, 2], the representation provides a simple explanation for the decoupling of change and spin degrees of freedom.

The p-particles associated with our new representation of the Lied and We solution are many–body collective modes which describe all the low–lying fluctuations of the model [1, 2]. Contrary to the quasiparticles of the Fermi liquid theory, which in the limit of vanishing interaction map onto real particles, the present class of p-particles cannot exist outside the many–body system for any value of the on–site repulsion U (including $U = 0$). Therefore, their coupling never vanish simultaneously [1, 2]. Nevertheless, they have many similarities with Fermi liquid quasiparticles [1, 2]. This explains our designation "Landau–Luttinger–Liquids" for fermionic Luttinger liquids of this kind.

---

* Alexander von Humboldt research fellow. Permanent address: University Evora, Ap.94, 7001 Evora CODEX, Portugal and CFMC–INIC Av. P.G. Pinto 2, 1699 Lisbona CODEX, Portugal

2. The New Representation

The 1–d Hubbard model reads

$$H = -t \sum_{j,\sigma}(C^+_{j\sigma}C_{j+1\sigma} + \text{h.c.}) + U \sum_j C^+_{j\uparrow}C_{j\uparrow}C^+_{j\downarrow}C_{j\downarrow} , \qquad (1)$$

where $C^+_{j\sigma}(C_{j\sigma})$ creates (annihilates) an electron with spin σ on site j ($j = 1,\ldots N_a$). As usual periodic boundary conditions are imposed $C_{N_a+1\sigma} \equiv C_{1\sigma}$. Equation (1) represents the interaction N electrons. In addition to the electronic density $n = N/N_a$, $0 \leq n \leq 1$ ($K_F = \pi/2n$), we use the ratio of the on–site repulsion U to the $U = 0$ bandwidth, $u = U/4t$, to describe the model parameter space. We introduce the function $K(q)$ and $S(p)$ defined by

$$\frac{dK(q)}{dq} = \frac{1}{2\pi\rho(K(q))}, \qquad \frac{dK^{-1}(k)}{dk} = 2\pi\rho(k) , \qquad (2)$$

$$\frac{dS(p)}{dp} = \frac{1}{2\pi\beta(S(p))}, \qquad \frac{dS^{-1}(s)}{ds} = 2\pi\beta(s) , \qquad (3)$$

where $K^{-1}(k)$ and $S^{-1}(s)$ are the inverse functions of $K(q)$ and $S(p)$, respectively, and $\rho(k)$ and $\beta(s)$ are the distributions appearing in the Lieb and Wu equations [2, 3] (we have replaced $\sigma(\Lambda)$ by $\beta(s) = u\sigma(us)$, $s = u^{-1}\Lambda$). In the case of the ground state, $K(q) = -K(-q)$, $S(p) = -S(-p)$ and $K(2K_F) = Q$, $K(\pi) = \pi$, $S(K_F) = \infty$, where Q is the cut off appearing in the Lieb and Wu equations [3]. These equivalent to

$$K(q) = q + \frac{1}{\pi}\int_{-K_0}^{K_0} dp' N_\downarrow(p') \tan^{-1}[S(p') - u^{-1}\sin K(q)] , \qquad (4)$$

$$p = \frac{1}{\pi}\int_{-\pi}^{\pi} dq' M_c(q') \tan^{-1}[S(p) - u^{-1}\sin K(q')]$$
$$- \frac{1}{\pi}\int_{-K_0}^{K_0} dp' N_\downarrow(p') \tan^{-1}[\frac{1}{2}(s(p) - S(p'))] , \qquad (5)$$

where K_0 is determined by the equation

$$K_0 = \frac{1}{2}\left[\int_{-pi}^{\pi} dq' M_c(q') - \int_{-K_0}^{K_0} dp' N_\downarrow(p')\right] , \qquad (6)$$

and is such that $S(K_0) = \infty$. For the ground state one obtains $K_0 = K_F$ and

$$M_c(q) = \theta(2K_F - |q|), \qquad N_\downarrow(p') = \theta(K_F - |p|) , \qquad (7)$$

which are the pseudo–momentum distributions of the charge and spin p-particles, that for the ground state have pseudo–Fermi surface at $q = \pm 2K_F$ and $p = \pm K_F$, respectively. As in the case of Landau Fermi liquid theory, these distributions are independent of the interaction. Moreover, (4–6) are valid for arbitrary choice of $M_c(q)$ and $N_\downarrow(p)$. In fact, in the thermodynamic limit, $N_a \to \infty$, there is a complete correspondence between the charge and spin pseudo–momentum distributions and the distributions of numbers I_j and J_α of reference [3]. The restrictions on these sets of numbers leads to a fermionic character for the charge and spin p-particles [2]. The quantum numbers which characterize these p-particles are the momentum $q(|q| \leq \pi)$ and charge e, and the momentum $p(|p| \leq K_0)$ and spin projection $-1/2$ (or $1/2$), respectively. The spin subsystem has a pseudo–Brillouin zone of width $2K_0$ ($2K_F$ for to ground state and low–lying spin excitations). In addition to $M_c(q)$ and $N_\downarrow(p)$ it is useful to introduce the related distributions $\widetilde{M}_c(k)(|k| \leq \pi)$ and $\widetilde{N}_\downarrow(s)(-\infty \leq s \leq \infty)$, so that

$$\widetilde{M}_c(K(q)) = M_c(q), \quad \widetilde{N}_\downarrow(S(p)) = N_\downarrow(p) . \qquad (8)$$

In the case of the ground state $\widetilde{M}_c(k)$ and $\widetilde{N}_\downarrow(s)$ are given by the following equations

$$\widetilde{M}_c(k) = \theta(Q - |k|), \quad \widetilde{N}_\downarrow(s) = 1 . \qquad (9)$$

The Lieb and Wu equations [3] can be generalized to

$$2\pi\rho(k) = 1 + \frac{u^{-1}}{\pi}\cos k \int_{-\infty}^{\infty} ds\, \widetilde{N}_\downarrow(s) \frac{2\pi\beta(s)}{1 + (s - u^{-1}\sin k)^2} \qquad (10)$$

$$\begin{aligned} 2\pi\beta(s) = & \frac{1}{\pi}\int_{-\pi}^{\pi} dk\, \widetilde{M}_c(k)\frac{2\pi\rho(k)}{1 + (s - u^{-1}\sin k)^2} \\ & - \frac{2}{\pi}\int_{-\infty}^{\infty} ds'\, \widetilde{N}_\downarrow(s') \frac{2\pi\beta(s')}{4 + (s - s')^2} . \end{aligned} \qquad (11)$$

The choice (9) leads to the usual form of the Lieb and Wu equations. The energy can be written as follows [2, 3]

$$\begin{aligned} E = & \frac{N_a}{2\pi}\int_{-\pi}^{\pi} dk\, \widetilde{M}_c(k) 2\pi\rho(k)[-2t\cos k] \\ = & \frac{N_a}{2\pi}\int_{-\pi}^{\pi} dq\, M_c(q)[-2t\cos K(q)] . \end{aligned} \qquad (12)$$

The combination of (4, 5,12) describes the interaction of the of the charge and spin p–particles. These p–particles can be identified with the "pseudo–Fermions" and the "antiholons and antispinons" considered in [6, 4], respectively.

The central equation of the new representation defines the energy E in terms of arbitrary fluctuations around the ground state. In order to derive it we introduce

$$M_c(q) = \theta(2K_F - |q|) + \delta_c(q), \quad N_\downarrow(p) = \theta(K_F - |p|) + \delta_\downarrow(p) , \qquad (13)$$

where $\delta_c(q)$ and $\delta_\downarrow(p)$ are small arbitrary fluctuations around the ground state pseudo-momentum distributions. Expansion of (4, 5, 12) around these distributions, (7), leads to the Landau energy

$$E = \sum_{n=0} E_n , \qquad (14)$$

where E_0 is the ground state energy, and the first and second terms of the r.h.s. of (14) are given by

$$E_1 = \frac{N_a}{2\pi} \int_{-\pi}^{\pi} dq\, \delta_c(q)\varepsilon_c(q) + \frac{N_a}{2\pi} \int_{-K_F}^{K_F} dp\, \delta_\downarrow(p)\varepsilon_\downarrow(p) , \qquad (15)$$

$$E_2 = \frac{N_a}{4\pi^2} \int_{-\pi}^{\pi} dq \int_{-\pi}^{\pi} dq'\, \delta_c(q)\delta_c(q')\frac{1}{2}f_{cc}(q,q')$$

$$+ \frac{N_a}{4\pi^2} \int_{-K_F}^{K_F} dp \int_{-K_F}^{K_F} dp'\, \delta_\downarrow(p)\delta_\downarrow(p')\frac{1}{2}f_{\downarrow\downarrow}(p,p')$$

$$+ \frac{N_a}{4\pi^2} \int_{-\pi}^{\pi} dq \int_{-K_F}^{K_F} dp\, \delta_c(q)\delta_\downarrow(p) f_{c\downarrow}(q,p) . \qquad (16)$$

As we will show in the following, particular choices of the fluctuations $\delta_c(q)$, $\delta_\downarrow(p)$ allow the exact calculation of physical quantities such as the magnetic susceptibility, the low temperature specific heat, etc. If we choose $\delta_c(q)$, $\delta_\downarrow(p)$ to obey the normalization conditions

$$\int_{-\pi}^{\pi} dq\, \delta_c(q) = \pm\frac{2\pi}{N_a}, \quad \int_{-K_F}^{K_F} dp\, \delta_\downarrow(p) = -\frac{2\pi}{N_a} , \qquad (17)$$

the excitation energy terms on the r.h.s. of (15, 16) define the spectra of the charge and spin p-particles in the same way as Fermi liquid theory. Similarly the charge-charge $f_{cc}(q,q')$, spin-spin $f_{\downarrow\downarrow}(p,p')$ and charge-spin $f_{c\downarrow}(q,p)$ f functions are associated with the two p-particle zero momentum transfer forward scattering (E_n with the n p-particle forward-scattering).

When the arbitrary fluctuations $\delta_c(q)$, $\delta_\downarrow(p)$ involve a number of p-particles $\ll N_a$, they are of the order of $1/N_a$. In the following we will show that all low-lying excited states of the model can be described by fluctuations of that order. This fact, together with the structure of the r.h.s. of (14-16), ex[plains the reason for the charge-spin decoupling which characterized the low-lying excitations of the model. In fact, as the contribution of E_n, $n \geq 2$, to the excitation energy is of the order of $(1/N_a)^{n-1}$ relative to E_1, $\sum_{n\geq 2} E_n$ can be neglected in the present thermodynamic limit. Nevertheless the amplitudes of the p-particles scattering play an important role in the spectral properties of the model [1, 2].

3. The P-Particles Spectra and Forward Scattering Amplitudes

A detailed derivation of the p-particles spectra, $\varepsilon_c(q), \varepsilon_\downarrow(p)$, and forward scattering amplitudes, $f_{cc}(q,q')$, $f_{\downarrow\downarrow}(p,p')$, $f_{c\downarrow}(q,p)$ is presented in reference [2]. The following expressions can be derived for $\varepsilon_c(q)$ and $\varepsilon_\downarrow(p)$

$$\varepsilon_c(q) = -2t\cos K(q) - 4t \int_0^\infty d\omega \, \frac{\Phi_1(\omega)\cos[\omega u^{-1}\sin K(q)]}{\omega(1+e^{2\omega})}, \quad (18)$$

$$\varepsilon_\downarrow(p) = -2t \int_0^\infty d\omega \, \frac{\Phi_1(\omega)\cos[\omega s(p)]}{\omega \cos h\omega}, \quad (19)$$

where $K(q)$ and $S(p)$ are the ground state solutions of (4, 5), which can be defined in terms of the inverse functions $K^{-1}(k)$ and $S^{-1}(s)$, respectively. These are given by

$$K^{-1}(k) = k + 2 \int_0^\infty d\omega \, \frac{\Phi_0(\omega)\sin(\omega u^{-1}\sin k)}{\omega(1+e^{2\omega})}, \quad (20)$$

$$S^{-1}(s) = \int_0^\infty d\omega \, \frac{\Phi_0(\omega)\sin(\omega s)}{\omega \cos h\omega}. \quad (21)$$

The functions $\Phi_0(\omega)$ and $\Phi_1(\omega)$ are solutions of related simple integral equations of common Kernel, so that

$$\Phi_j(\omega) = F^{(j)}(\omega) + \int_{-\infty}^\infty d\omega' \Gamma(\omega,\omega')\Phi_j(\omega'), \quad j=0,1, \quad (22)$$

where the free terms and the kernel are defined as

$$F^{(0)}(\omega) = \frac{1}{2\pi} \int_{-Q}^Q dk \, \cos(\omega u^{-1}\sin k), \quad (23)$$

$$F^{(1)}(\omega) = \frac{d}{d(\omega u^{-1})} F^{(0)}(\omega) = \frac{1}{2\pi} \int_{-Q}^Q dk \, \sin k \sin(\omega u^{-1}\sin k), \quad (24)$$

$$\Gamma(\omega,\omega') = \frac{\sin[(\omega-\omega')u^{-1}\sin Q]}{\pi(\omega-\omega')} \frac{1}{1+e^{2|\omega'|}}. \quad (25)$$

For the particular case of the half-filled band, $n=1$, we obtain $\Phi_j(\omega) = J_j(\omega u^{-1})$, $j=0,1$, where J_j, are Bessel functions.

The spectra (18, 19) have the following asymptotic expressions

$$\varepsilon_c(q) = \begin{cases} -4t\cos\left(\frac{q}{2}\right) + 2t\cos\left(\frac{\pi}{2}n\right), \\ \quad 0 \leq |q| \leq 2K_F, \quad U=0, \\ -2t\cos(|q|) - \frac{\pi}{2}n), \\ \quad 2K_F \leq |q| \leq \pi, \quad U=0, \\ -2t\cos q - \frac{t^2}{U}8n\ln(2)\left[\sin^2 q + \frac{1}{2}\left(1 - \frac{\sin(2\pi n)}{2\pi n}\right)\right], \\ \quad U \gg t, \end{cases} \quad (26)$$

$$\varepsilon_\downarrow(p) = \begin{cases} -2t[\cos p - \cos\left(\frac{\pi}{2}n\right)], & U = 0, \\ -\frac{t^2}{U}2\pi n\left[1 - \frac{\sin(2\pi n)}{2\pi n}\right]\cos\left(\frac{p}{n}\right), & U \gg t \end{cases} \tag{27}$$

The velocities

$$v_c(q) = \frac{d\varepsilon_c(q)}{dq}, \quad v_\downarrow(p) = \frac{d\varepsilon_\downarrow(p)}{dp}, \tag{28}$$

which are obtained by differentiation of the r.h.s. of (18, 19), play an important role in determining the physics of the model. Their asymptotic behavior is

$$v_c(q) = \begin{cases} 2t\sin\left(\frac{q}{2}\right), & 0 \leq |q| \leq 2K_F, \ U = 0, \\ (\text{sgn}q)2t\sin\left(|q| - \frac{\pi}{2}n\right), & 2K_F \leq |q| \leq \pi, \ U = 0, \\ 2t\sin q - \frac{t^2}{U}8n\ln(2)\sin 2q, & U \gg t, \end{cases} \tag{29}$$

$$v_\downarrow(p) = \begin{cases} 2t\sin p, & U = 0, \\ \frac{t^2}{U}2\pi\left[1 - \frac{\sin(2\pi n)}{2\pi n}\right]\sin\left(\frac{p}{n}\right), & U \gg t \end{cases} \tag{30}$$

The full description of the low–lying excitations of the model also involves a band $\varepsilon_\uparrow(p)$ for up–spin p-particles ($\varepsilon_\uparrow(p) = \varepsilon_\downarrow(p)$ in the ground state) and an upper "hole" charge band

$$\varepsilon_c^h(q) = U - \varepsilon_c(q), \tag{31}$$

which is separated from the bottom of the lower $\varepsilon_c(q)$ band by the correlation gap [1, 2]

$$\Delta = \varepsilon_c^h(\pi) - \varepsilon_c(\pi) = U - 4t + 8t\int_0^\infty d\omega \frac{\Phi_1(\omega)}{\omega(1 + e^{2\omega})}. \tag{32}$$

For the half–filled band case the r.h.s. of (32) coincides with the well known gap for charge excitations [3, 7, 8].

For $U/t \to \infty$, $\varepsilon_c(q) = -2t\cos q$ and $\varepsilon_\downarrow(p)$ is dispersionless, $\varepsilon_\downarrow(p) = 0$. In that limit the change and spin p-particles are the spinless fermions and the localized spins of the "squeezed" Heisenberg model of reference [9], respectively. For finite and vanishing U, $\varepsilon_\downarrow(p)$ is negative except at the pseudo–Fermi points $p = \pm K_F$ where it vanishes. Starting from the $U \to \infty$ limit, the effect of decreasing U is to increase the bandwidth of $\varepsilon_\downarrow(p)$ and renormalize $\varepsilon_c(q)$ without changing its bandwidth $4t$. The change velocity at $q = \pi$, $v_c(\pi)$, has a non–singular behaviour at $U = 0$ for all densities. In fact, while $v_c(\pi) = 0$ for $U > 0$, at $U = 0$ it becomes $v_c(\pi) = 2t\sin K_F$. For the half–filled band case that reflects the Mott transition. This non–singular behaviour, which is also present at the other densities, reflects the appearance of the correlation gap, (32), in the limits of the Brillouin zone of the charge p-particles. Moreover, the pseudo–Fermi velocities, $v_c(2K_F), v_\downarrow(K_F)$, which play an important role in the following, change from $v_c(2K_F) = V_\downarrow(K_F) = v_F = 2t\sin K_F$ at $U = 0$, to $v_c(2K_F) = 2t\sin 2K_F$, $v_\downarrow(K_F) = 0$ at $U/t \to \infty$.

The following expressions can be derived for the forward scattering amplitudes $f_{cc}(q,q')$, $f_{\downarrow\downarrow}(p,p')$ and $f_{c\downarrow}(q,p)$ [2]:

$$f_{cc}(q,q') = 2\pi[v_c(q)G_{cc}(q,q') + v_c(q')G_{cc}(q',q)]$$
$$+ 2\pi v_c(2K_F)\sum_{j=\pm 1} G_{cc}(2K_Fj,q)G_{cc}(2K_Fj,q')$$
$$+ 2\pi v_\downarrow(K_F)\sum_{j=\pm 1} G_{\downarrow c}K_Fj,q)G_{\downarrow c}(K_Fj,q') , \quad (33)$$

$$f_{\downarrow\downarrow}(p,p') = 2\pi[v_\downarrow(p)G_{\downarrow\downarrow}(p,p') + v_\downarrow(p')G_{\downarrow\downarrow}(p',p)]$$
$$+ 2\pi v_\downarrow(K_F)\sum_{j=\pm 1} G_{\downarrow\downarrow}(K_Fj,p)G_{\downarrow\downarrow}(K_Fj,p')$$
$$+ 2\pi v_c(2K_F)\sum_{j=\pm 1} G_{c\downarrow}(2K_Fj,p)G_{c\downarrow}(2K_Fj,p') , \quad (34)$$

$$f_{c\downarrow}(q,p) = 2\pi[v_c(q)G_{c\downarrow}(q,p) + v_\downarrow(p)G_{\downarrow c}(p,q)]$$
$$+ 2\pi v_c(2K_F)\sum_{j=\pm 1} G_{cc}(2K_Fj,q)G_{c\downarrow}(2K_Fj,p)$$
$$+ 2\pi v_\downarrow(K_F)\sum_{j=\pm 1} G_{\downarrow\downarrow}K_Fj,p)G_{\downarrow c}(K_Fj,q) , \quad (35)$$

where the functions $G_{cc}(q,q')$, $G_{\downarrow\downarrow}(p,p')$, $G_{c\downarrow}(q,p)$ and $G_{\downarrow c}(p,q)$ are related with phase shifts of the p-particles [2]. For $U=0$ and $U/t \to \infty$ they are given by the following expressions

$$G_{cc}(q,q') = \begin{cases} [(\text{sgn}(\overline{q}-\overline{q}'))(1-\delta_{\overline{q},\overline{q}'})]\frac{1}{2}, & U=0, \\ 0, & U/t \to \infty \end{cases}, \quad (36)$$

$$G_{\downarrow\downarrow}(p,p') = \begin{cases} 0, & U=0, \\ \frac{1}{2\pi}[\psi_{p,p'}], & U/t \to \infty \end{cases}, \quad (37)$$

$$G_{c\downarrow}(q,p) = \begin{cases} -[G_{\downarrow c}(p,q)], & U=0, \\ \frac{p}{2\pi n}, & U/t \to \infty \end{cases}, \quad (38)$$

$$G_{\downarrow c}(p,q) = \begin{cases} [\text{sgn}(p-q/2)](1-\delta_{p,q/2})(1-\delta_{p,\pm K_F}\delta_{q,\text{sgn}p\pi}\frac{1}{2}), & U=0, \\ -\frac{p}{2\pi n}, & U \to \infty \end{cases}, \quad (39)$$

where

$$\overline{q} = \begin{cases} q, & 0 \leq |q| \leq \frac{\pi}{2} + K_F, \\ \text{sgn}q(\pi+2K_F) - q, & \frac{\pi}{2} + K_F \leq |q| \leq \pi \end{cases}, \quad (40)$$

and, $-\pi/2 \leq \psi_{p,p'} \leq \pi/2$, reads

$$\psi_{p,p'} = \frac{1}{\pi}\int_{-\infty}^{\infty} dx \frac{\tan^{-1}[\sin h(\frac{\pi}{2}x + \sin h^{-1}(\tan(\frac{p}{n})) - \sin h^{-1}(\tan(\frac{p'}{n})))]}{1+x^2}. \quad (41)$$

The f functions, (33–35), never vanish simultaneously. For example for $U/t \to \infty$, $f_{cc}(q,q') = 0$, but $f_{\downarrow\downarrow}(p,p')$ and $f_{c\downarrow}(q,p)$ remain finite. In this case we have have non–interacting spinless fermions, whose hopping is "seen" by the localized spins of the "sqeezed" Heisenberg model. In can be shown that these $U \to \infty$ p-particle scattering smear out the the $2K_F$ anomaly of the electronic momentum distribution [2, 9]. Although the f functions do not lead to bound states in the 1–d Hubbard model, they regulate its spectral properties. In references [1], [2] the relation of the amplitude $\mid f_{c\downarrow}(q,p) \mid$ with the ground state electronic spectral function $A(k,\omega)$ is studied. Moreover, similar to the case of the simpler Luttinger liquids [5], all the critical exponents of the 1–d Hubbard model can be expressed in terms of the renormalized velocities and couplings provided by the new exact representation [2].

4. Low Energy Physics of the Model

Besides determining the non–trivial spectral properties of the model [1, 2], the charge and spin p-particles describe the thermodynamics of the low–lying excited states of the system. The charge and spin excitations of the model, which were studied by different authors using more or less involved calculations, can be simply expressed in terms of the spectra $\varepsilon_c(q), \varepsilon_c^h(q)$ and $\varepsilon_\downarrow(p)$ (or $\varepsilon_\uparrow(p)$). The spectra of the charge "particle–hole" gapless [7,10] and across–gap [8] excitations, as well as of the triplet "spin–wave" excitations [7, 10], can be written as

$$E_{p-h}(k) : E_{p-h} = \varepsilon_c(q_1) - \varepsilon_c(q_0), \quad k = q_1 - q_0, \tag{42}$$

$$E_{p-h}^\Delta(k) : E_{p-h}^\Delta = \varepsilon_c^h(q_1) - \varepsilon_c(q_0), \quad k = q_1 - q_0, \tag{43}$$

$$E_{sw}(k) : E_{sw} = -\varepsilon_\downarrow(p_0) - \varepsilon_\downarrow(p_1), \quad k = 2K_F - p_0 - p_1, \tag{44}$$

respectively. (42–44) show that different configurations of the occupations of $M_c(q)$ and $N_\downarrow(p)$ generate all the low–lying excited states of the model. Our representation enables the characterization of all the degenerate spin excitations (case of the singlet states is treated in [2]).

In addition to the excitations, (42–44), further excitations can be generated by removing an increasing number of p-particles ($\ll N_a$) out of the pseudo–Fermi seas.

In the thermodynamic limit we can study low temperature thermodynamics [2]. For temperatures which $K_B T \ll [\varepsilon_c^h(\pi) - \varepsilon_c(2K_F)]$ this is achieved by replacing $M_c(q)$ and $N_\downarrow^c(p)$ by Fermi distributions

$$M_c(q) = \frac{1}{1 + e^{[\varepsilon_c(q) - \varepsilon_c(2K_F)]\beta}}, \tag{45}$$

$$N_\downarrow(p) = \frac{1}{1 + e^{\varepsilon_\downarrow(p)\beta}}, \tag{46}$$

where $\varepsilon_c(q)$ and $\varepsilon_\downarrow(p)$ are ground state spectra, (18, 19) (note that $\varepsilon_\downarrow(K_F) = 0$). The low temperature specific heat, which is linear in T for all parameter space [1, 2], can be simply obtained by introducing the following small thermal fluctuations

$$\delta_c(q) = M_c(q) - \theta(2K_F - (q)), \quad n < 1; \quad \text{all} \quad n, \quad U = 0,$$
$$0, \qquad\qquad\qquad\qquad\qquad n = 1, \quad U > 0, \quad (47)$$
$$\delta_\downarrow(p) = N_\downarrow(p) - \theta(K_F - (p)), \quad (48)$$

where $M_c(q)$ and $N_\downarrow(p)$ are given in (45, 46). The corresponding energy reals

$$E = E_0 + \frac{N_a}{2\pi} \int_{-\pi}^{\pi} dq \delta_c(q)(\varepsilon_c(q) - \varepsilon_c(2K_F))$$
$$+ 2\frac{N_a}{2\pi} \int_{-K_F}^{K_F} dp \delta_\downarrow(p)\varepsilon_\downarrow(p), \quad (49)$$

where the contributions associated with factor 2 will be discussed in reference [2]. Linearizing the charge and spin spectra around the pseudo–Fermi points $\pm 2K_F$ and $\pm K_F$, respectively, leads to the following exact expression for the low temperature specific heat, which is valid for the whole parameter space $U \geq 0$, $0 \leq n \leq 1$

$$c_v = \begin{cases} [\pi N_a K_B{}^2/3 v_\downarrow(K_F)]T, \\ \qquad n = 1, \quad U > 0, \\ [(\pi N_a K_B{}^2/3)(V_c(2K_F)^{-1} + V_\downarrow(K_F)^{-1}], \\ \qquad n > 1, \quad U > 0; \quad \text{all} \quad n, \quad \text{if} \quad U = 0. \end{cases} \quad (50)$$

This expression leads to the correct $U = 0$ low temperature specific heat. The non–analytic behaviour at $U = 0$ of the half–filled band expression reflects the Mott transition [2]. For $n = 1$ the large U first order t^2/U expansion of the r.h.s. of (50) is consistent with the result obtained by Takahashi [11] for the Heisenberg chain.

The spin magnetic susceptibility can also be simply derived by adding a small magnetic field H and introducing the following spin fluctuation $(K_{F\downarrow} < K_{F\uparrow})$ $\delta_\downarrow(p) = \theta(K_{F\downarrow} - |p|) - \theta(K_F - |p|)$. The obtained energy reads

$$E = E_0 + \frac{N_a}{2\pi} \int_{-K_F}^{K_F} dp \delta_\downarrow(p)\varepsilon_\downarrow(p)$$
$$+ \frac{N_a}{4\pi^2} \int_{-K_F}^{K_F} dp \int_{-K_F}^{K_F} dp' \delta_\downarrow(p)\delta_\downarrow(p') \frac{1}{2} f_{\downarrow\downarrow}(p, p') \quad (51)$$
$$- \frac{2}{\pi} N_a \mu_0 H(K_F - K_{F\downarrow}).$$

Minimization of the r.h.s. of (51) with respect to the magnetization leads, after linearization of $\varepsilon_\downarrow(p)$ around $p = \pm K_F$, to the following exact expression for the magnetic spin susceptibility

$$\chi = \frac{2\mu_0^2}{\pi V_\downarrow(K_F)} \ . \tag{52}$$

For $U = 0$ the r.h.s. of (52) reduces to the Pauli susceptibility. For $U \gg t$ it results in the asymptotic expansion derived by Shiba [12]. Moreover, for the particular case of the half–filled band, it agrees with the expression obtained by Takahashi [2, 13], and for $U \gg t$ is consistent with Griffiths result for the Heisenberg chain [14].

As in Fermi liquid theory, the form of the low temperature specific heat and magnetic susceptibility expression is the same for all values of the electronic interactions. Nevertheless, contrary to Fermi liquid theory, the interactions renormalize two pseudo–Fermi velocities $V_c(2K_F)$ and $V_\downarrow(K_F)$, corresponding to the charge and spin p-particles.

5. Concluding Remarks

In this paper we used a new representation of the Bethe ansatz solution of the 1–d Hubbard Hamiltonian to study the low energy physics of the model in terms of interacting charge and spin p-particles. Although these p-particles are many–body collective modes which cannot exist outside the system, they are the real elementary entities which control both the non–trivial spectral properties of this Landau–Luttinger liquid [1, 2] and the low–lying excitations of the model. Moreover they can couple to external fields, explaining for example the $4K_F$ (charge) and $2K_F$ (spin) diffuse x–ray scattering observed in quasi–one dimensional conductors [1, 15]. In addition to their relevance to these compounds, the new picture arising from the interaction of these p-particles may offer insight into the physics of higher dimensional systems [4]. In fact, contrary to the 1–d model, where the interaction of the new Landau p-particles does not lead to bound states [2], their interaction could possibly produce bound states in these systems, introducing a mechanism for high T_c superconductivity.

Acknowledgements We thank P.A. Maksym for revising the manuscript and P. Horsch and W. Stephan for stimulating discussions. J.C. was supported by the Alexander von Humboldt–Stiftung.

References

1. J. Carmelo and A.A. Ovchinnikov: Preprint (1990)
2. J. Carmelo, P. Horsch and A.A. Ovchinnikov: submitted to Phys. Rev. **B** (1990)
3. E.H. Lieb and F.Y. Wu: Phys. Rev. Lett. **20** 1445 (1968)
4. P.W. Anderson and Y. Ren: Princeton University, Preprint (1990)
5. F.D.M. Haldane: Phys. Rev. Lett. **45** 1358 (1980); Physics Letters **81A** 153 (1981); J. Phys. **C 14** 2585 (1981)

6. J. Carmelo and D. Baeriswyl: Int. J. Mod. Phys. **1** 1013 (1988)
7. A.A. Ovchinnikov: Zh. Eksp. Teor. Fiz. **57** 2137 (1969) (Sov. Phys. JETP **30** 1160 (1970)
8. F. Woynarovich: J. Phys. **C 15** 85 (1982)
9. M. Ogata and H. Shida: Phys. Rev. **B 41** 2326 (1990)
10. C.F. Coll III: Phys. Rev. **B 9** 2150 (1974); T.C. Choy and W. Young, Y.: Phys. **C 15** 521 (1982)
11. M. Takahashi: Prog. Theor. Phys. **47** 69 (1972); **50** 1519 (1973)
12. H. Shiba: Phys. Rev. **B 6** 930 (1972)
13. M. Takahashi: Prog. Theor.Phys. **42** 1098 (1969); **43** 1619 (1970)
14. Robert B. Griffiths: Phys. Rev. **133** A 768 (1964)
15. J.P. Pouget, S.K. Khanna, F. Denoyer and R. Comes: Phys. Rev. Lett. **37** 437 (1976)

Mean–Field Study of Possible Electronic Pairings in the CuO Plane of HTSO

A.A. Ovchinnikov and M.Ya. Ovchinnikova

Institute of Chemical Physics, Kosygin St. 4, Moscow, 117 334, USSR

Since the discovery of the high T_c superconductivity [1] a great variety of different pairing mechanisms have been proposed, in particular those caused by local short range attractive interaction [2]. The aim of this paper is to study the possible origin of this attraction. From this point all local interactions of the electronic $d_{x^2-y^2}, p_x, p_y$ orbits of adjacent atoms in CuO_2 plane are classified and analyzed in terms of the MF theory for the upper band in the known three band model [3–7] of CuO_2 plane in assumption of frozen two lower bands. The main parameters of model have been discussed in [8–12]. Preliminary the Hartree–Fock approximation (HFA) is used to obtain the upper band parameters varying with doping.

The MF theory in the band approach is used unlike the MF studies [13–15] of electron pairing in the Mott–Hubbard approach. A comparative discussion of both approaches are given in [16]. Present study takes into account more detailed picture of interactions in system (not only on–site and inter–site parameters) and it is similar to study [17] of antiferromagnet (AF) pairing. In spite of the limited accuracy of MF theory the extending and the MF classification of all types of interactions in problem and a numerical study of models may be instructive.

Contrary conventional opinion [2] the hope on superconductivity (SC) may be connected with namely the correlated hopping interaction among all considered local interactions. Note that here we deal with the correlated hopping between the p and d orbits instead of similar effective hopping interaction between elementary sites. Moreover the existence of SC depends crucially on $E_d = \varepsilon_d - \varepsilon_p$, i.e., on difference in renormalized energies of p, d orbits. If $E_d > 0$ the holes populate preferentially the Cu sites and the antiferromagnet state (AF) spreads over large range of doping and suppresses the SC. In case of $E_d < 0$ the AF correlations of spins on Cu centers weaken and these is a large region of doping when superconductivity can exist for our models at sufficiently large K_d, K_p. It may be of s- or d-type depending on K_d/K_p.

The known model [3–7] for the bands is spanned on the basis of $Cu(d_{x^2-y^2}), O(p_x), O(p_y)$ orbits which correspond to the electron creation operators $d^+_{n\sigma}, x^+_{n\sigma}, y^+_{n\sigma}$. The Hamiltonian of the model is

$$H = H_0(t_0, \varepsilon_p^0, \varepsilon_d^0) + V_U + V_Q + V_J + V_K ,\qquad(1)$$

where H_0 is the one–electron "zero" Hamiltonian

$$\begin{aligned}H_0 = t_0 \sum_\sigma \sum_{(nm)} \zeta_{nm}(d_{n\sigma}^+ x_{m\sigma} - d_{n\sigma}^+ y_{m\sigma} + \text{h.c.})\\ + \sum_{n\sigma} \varepsilon_d d_{n\sigma}^+ d_{n\sigma} + \sum_{n\sigma} \varepsilon_p(x_{n\sigma}^+ x_{n\sigma} + y_{n\sigma}^+ y_{n\sigma}) .\end{aligned}\qquad(2)$$

The summation of (m, n) is taken over nearest–neighbor Cu and O centers, i.e., $m - n$ is equal to zero or e_x or e_y, and $\zeta_{mn} = (-1)^{|n-m|}$ if the natural signs of orbits are chosen [18]; $t_0, \varepsilon_d^0, \varepsilon_p^0$ are the "zero" $p-d$ transfer integral and orbital energies; V_U contains the one–center Coulomb integrals U_d, U_p on atoms Cu and O; V_Q, V_J correspond to the Coulomb and exchange integrals Q and J for the nearest–neighbor p and d orbits, V_K describes the correlated hopping and provides dependence of actual $p-d$ transfer integral on occupancy of these orbits. V_K have a from

$$\begin{aligned}V_K = -K_d \sum_{(nm)\sigma} \zeta_{nm}\{[d_{n\sigma}^+ d_{n-\sigma}^+ x_{m-\sigma} d_{n\sigma} + \text{h.c.}] - [\ldots]_y\}\\ - K_p \sum_{(nm)\sigma} \zeta_{nm}\{[x_{n\sigma}^+ x_{n-\sigma}^+ d_{m-\sigma} x_{n\sigma} + \text{h.c.}] - [\ldots]_y\} .\end{aligned}\qquad(3)$$

Here $[\ldots]_y$ denotes the same as previous brackets for y operators. In limit of small overlapping of p, d orbits the parameter K_d is associated with integral $K_d = -\langle d_{n\sigma}(1) d_{n-\sigma}(2) | H - H_{HF} | d_{n\sigma}(1) x_{n-\sigma}(2) \rangle$ over coordinates of two electrons and K_p is the same with $d \leftrightarrows p$. If the basis orbits are considered firm i.e., invariable in course of their occupation, then one excepts $K_d, K_p > 0$ as well as $J > 0$ for the sign of orbits chosen. Electronic representation (instead of the hole one) is used here.

Ordinary HFA to (1) leads to the Hamiltonian which coincides with $H_0(t, \varepsilon_p, \varepsilon_d)$ but with replacement of parameters $t_0, \varepsilon_d^0, \varepsilon_p^0$ by renormalized values

$$\begin{aligned}\varepsilon_d &= \varepsilon_d^0 + U_d \rho_d + 4(2Q + J)\rho_d - 8K_d \rho_{pd} ,\\ \varepsilon_p &= \varepsilon_p^0 + U_p \rho_p + 4(2Q + J)\rho_p - 4K_p \rho_{pd} ,\\ t &= t_0 - (Q + 2J)\rho_{pd} - K_d \rho_d - K_p \rho_p ,\end{aligned}\qquad(4)$$

where $\rho_d = \langle d_{n\sigma}^+ d_{n\sigma}\rangle$, $\rho_p = \langle x_{n\sigma}^+ x_{n\sigma}\rangle$, $\rho_{pd} = \zeta_{nm}\langle x_{m\sigma}^+ d_{n\sigma}\rangle$ are average occupations of p, d orbits and the bond order value for definite spin projection σ. The energies of three bands (bonding, antibonding and dispersionless nonbonding ones) and the annihilation operators of the band states are

$$\varepsilon_{1(3)} = \bar{\varepsilon} \pm \sqrt{D^2 + W^2};\quad \varepsilon_2 = \varepsilon_p ,\qquad(5)$$

$$a^\lambda_{k\sigma} = N^{-1/2} \sum_{n,\sigma} e^{-ikn}(d_{n\sigma}, x_{n\sigma} y_{n\sigma})_j A_{j\lambda}, \quad j,\lambda = 1,2,3 . \tag{6}$$

Here the upper band coefficients and parameters are

$$A_{j1} = (C, ie^{-ik_x/2} S s_x, -ie^{-ik_y/2} S s_y)_j,$$

$$\bar{\varepsilon} = (\varepsilon_p + \varepsilon_d)/2, \quad D = (\varepsilon_p - \varepsilon_d)/2, \quad W = 2tw, \quad w = \sqrt{s_x^2 + s_y^2}, \tag{7}$$

$$s_x = \sin(k_x/2), \quad s_y = \sin(k_y/2), \quad C = \cos\theta,$$

$$S = \sin\theta/w, \quad \tang 2\theta = -W/D .$$

If the bands 2, 3 are completely filled and only the upper band is partly occupied then the self–consisting HF equations are

$$\rho_d = 1 - (1/N) \sum_k |A_{11}|^2 (1 - f_1) = n(\mu)/2 - 2\rho_p ,$$

$$\rho_{pd} = -(1/N) \sum_k \mathrm{Re}(A^* A_1)(1 - f_1) , \tag{8}$$

$$n(\mu) = 6 - (2/N) \sum_k (1 - f_1) = N_h + 5 ,$$

where n is a number of electrons on one site and N_h is the hole concentration associated with doping; f_1 is the Fermi distribution function in upper band and $\sum$ is a sum over k. All HF calculations have been done at zero temperature.

In representation of HF states (6) Hamiltonian (1) includes interactions between states inside each band as well as those with and between states of different bands. Main approximation of the present study consists in retaining all interactions inside the upper band and neglecting those with and between states of lower bands which assume to be frozen. In such approximation the interaction depends only on the creation operators $a^+_{k\sigma}$ of upper band and takes a from

$$V = 1/2 \sum V(k_1\sigma_1 \ldots k_4\sigma_4) a^+_{k_1\sigma_1} a^+_{k_2\sigma_2} a_{k_3\sigma_3} a_{k_4\sigma_4} \delta(k_1 + k_2 + k_3 + k_4) , \tag{9}$$

where $\sum$ is sum over $k_j\sigma_j$ and matrix elements in (9) are calculated with use of expansion (6).

Now the most general linearized Hamiltonian of the upper band contains three types of terms with large phase volume, namely those with

$$(k_1 k_2 k_3 k_4) = \{(k, -\tilde{k}', k', -\tilde{k}), (k, -\tilde{k}', -\tilde{k} - k')\} \tag{10}$$
$$\text{or} \quad (k, -k, -k', k') \quad \text{or} \quad (k, \tilde{k}, \tilde{k}', k')$$

in addition to usual HF terms. Here vector k is defined by

$$\tilde{k} = e_x(\pi k_x/|k_x| - k_x) + e_y(\pi k_y/|k_y| - k_y) . \tag{11}$$

Variants (10) refer to the electron–hole pairing $(k, -\tilde{k})$, the electron–electron pairing $(k, -k)$ or the same with $(k, \tilde{k})$. These three types of ordering correspond to the normal state, to superconducting one or to unstable state with the nonzero alternating anomalous averages. Such correspondence is found by study of response of different model systems on the long-wave electromagnetic field [18]. Therefore, we study only two first types of pairing from (10). Thus obtained model interaction is

$$V_{\text{mod}} = \sum \Delta_\mu^\nu \Delta_\mu^\nu / \kappa_\mu^\nu + \sum \Gamma_{SM}^\nu \Gamma_{SM}^\nu / \gamma_S^\nu, \qquad (12)$$

$$\Delta_\mu^\nu = \kappa_\mu^\nu \sum{}' \varphi_\mu^\nu(k) t_\mu(k), \quad t_\mu(k) = \sum{}'' (\sigma_\mu)_{\sigma\sigma'} a_{k\sigma}^+ a_{-\tilde{k}\sigma'}, \qquad (13)$$

$$\Gamma_{SM}^\nu = \gamma_{SM}^\nu \sum{}' g_S^\nu(k) r_{SM}^\nu(k), \quad r_{SM}^\nu(k) = \sum{}'' C_{\sigma\sigma'}^{SM}, a_{-k\sigma}, a_{k\sigma}. \qquad (14)$$

Here σ_μ, $\mu = 0, 1, 2, 3$, are the Pauli matrices; $C_{\sigma\sigma'}^{SM}$ are Clebsh–Gordon coefficients for spins 1/2; index ν numerates all possible operators Δ, Γ; $\sum$, $\sum'$, $\sum''$ are the sums over μ, ν, S, M or over k or over σ, σ' correspondingly. All constants κ, γ and weight functions φ, g have been classified [18] and expressed in terms of the expansion coefficients of the upper band states (6) and of the parameters of original Hamiltonian (1). The reducing of V_{mod} to separable form (12) follows from the summation rules for σ_μ and $C_{\sigma\sigma'}^{SM}$, after transformation to symmetric weight functions $\varphi_\mu^\nu(-\tilde{k}) = [\varphi_\mu^\nu(k)]^*$, $g_S^\nu(-k) = (-1)^S g_S^\nu(k)$ [18]. Classification of all constants κ^ν, γ^ν, and weight functions φ^ν, g^ν and their expressions via the original parameters of Hamiltonian (1) are given elsewhere [18], as well as the contributions of each of the considered local interactions $(U_d, U_p, Q, J, K_d, K_p)$ to the interaction constants. For SC pairing the negative effective constant (i.e., the energy gain in course of SC ordering) is possible if the correlated hopping integrals are sufficiently large and positive. Note that interaction V_k gives rise to terms in V_{mod} with both the negative and positive constants γ, but for the partly filled antibonding band the first term prevails.

First term in (12) refers to possible SDW with polarization μ ($\mu = 1, 2, 3$) if the corresponding real order parameters (OP) $\underline{\Delta}_\mu^\nu = \langle \Delta_\mu^\nu \rangle$ are nonzero. At $\mu = 0$ the real OP $\underline{\Delta}_0^\nu$ corresponds to CDW with doubling volume of elementary site. Depending on symmetry of function φ in (13) the OP $\underline{\Delta}$ means the alternation of the spin density ($\mu \neq 0$) or the charge density ($\mu = 0$) on cooper ions or on oxygen ions or on the $p-d$ bonds or corresponds to state of orbital antiferromagnet with the charge (or spin) currents of some symmetry. The second term in (10) is responsible for the SC pairing in singlet ($S = 0$) or triplet ($S = 1$) states if the corresponding order parameters obey $\underline{\Gamma}_{SM}^\nu = \langle \Gamma_{SM}^\nu \rangle \neq 0$. Symmetry of function $g(k)$ in (13) determines the s, p, or d type of pairing.

A Hamiltonian for the upper band $\varepsilon(k) = \varepsilon_1(k)$ in (6) together with retained model interaction (11)

$$H = H_1 + V_{\text{mod}}, \quad H_1 = \sum [\varepsilon(k) - \mu] a_{k\sigma}^+ a_{k\sigma} \quad (15)$$

can be transformed directly to corresponding linearized Hamiltonian of general form (constant C_L is given in [18])

$$H_L = H_1(\varepsilon_p, \varepsilon_d, t) + 2 \sum_{\nu\mu} \underline{\Delta}_\mu^\nu \Delta_\mu^\nu / \kappa_\mu^\nu + \sum_{\nu SM} (\underline{\Gamma}_{SM}^{+\nu} \Gamma_S^\nu M + \text{h.c.})/\gamma_S^\nu + c_L \; . \quad (16)$$

Here $\underline{\Delta}_\mu^\nu = \langle \Delta_\mu^\nu \rangle$ are set of the real OP corresponding to averages of Hermitian operators Δ and $\underline{\Gamma}_S^\nu M = \langle \Gamma_{SM}^\nu \rangle$ are generally complex OP.

According to standard method the annihilation operators $\beta_\lambda(k)$ of quasiparticles obey equations $[\beta_\lambda, H_L] = -E_\lambda \beta_\lambda$ and are constructed as a linear combinations of operators $a_{k\sigma}^+, a_{k\sigma}$. The number of terms in expansion depends on number of operators Δ, Γ in H_L. For simplicity we consider the situation with one type of the SDW polarization $\mu = 3$ and with one projection $M = 0$ of anomalous OP in (16). Commutation of H_L with $a_{k\uparrow}^+$, for example, generates four operators $\{a_{k\uparrow} \; a_{-\widetilde{k}\uparrow} \; a_{-k\downarrow}^+ \; a_{\widetilde{k}\downarrow}^+\}$ which may serve as a basis set for expansion of $\beta_\lambda(k), \lambda = 1 \div 4$. For uniformity we call these basis operators as new operators $b_j(k)$. Then

$$\beta_\lambda(k) = \sum_j b_j(k) U_{j\lambda}(k), \quad \beta_\lambda^+(k) = \sum_j (U^+)_{\lambda j} b_j^+ \quad (18)$$

$$\{b_1(k), b_2(k), b_3(k), b_4(k)\} = \{a_{k\uparrow} \; a_{-\widetilde{k}\uparrow} \; a_{-k\downarrow}^+ \; a_{\widetilde{k}\downarrow}^+\}, \quad k < F \; . \quad (19)$$

Full set of such independent operators can be obtained if vector k varies in the half of all phase volume with taking into account the spin doubling of original operators $a_{k\sigma}$. For instance let k varies in the $2D$ region F corresponding to filled states in the half populated original band. Then H_L takes a from

$$H_L = \sum_{ij=1}^4 \sum_{k<F} L_{ji} b^+(k) b_j(k) + C_L \; . \quad (20)$$

The equations for Hermitian matrix $L_{ij}(k)$ via the weight functions and OP are given elsewhere [18]. The quasiparticle spectrum E_λ and the unitary matrix $U_{j\lambda}$ of coefficients in (18) are determined by equations

$$L_{ij} U_{j\lambda}(k) = -U_{j\lambda}(k) E_\lambda(k) \quad (21)$$

and the averages of any product of operators are

$$\langle b_i^+(k) b_j(k') \rangle = \delta_{kk'} U_{i\lambda} f_\lambda U_{\lambda j}^+; \quad \langle b_i^+(k) b_j^+(k') \rangle = \langle b_i(k) b_j(k') \rangle = 0 \; , \quad (22)$$

$$F_\lambda = \{1 - \exp[-\beta E_\lambda(k)]\}^{-1} \; . \quad (23)$$

Similar averaging of operators Δ_0^ν, Δ_a^ν, Γ_{SO}^ν leads to following selfconsistent equations

$$\Delta_\mu^\nu = \kappa_\mu^\nu N^{-1} \sum \mathrm{Re}\{\varphi_\mu^\nu(k)T_{12} - (-1)^\mu \varphi_\mu^\nu(\tilde{k})T_{34}\}, \quad \mu = 0, 3, \qquad (24)$$

$$\Gamma_{SO}^\nu = -\gamma_S^\nu N^{-1} \sum \{g_\mu^\nu(k)T_{31} + g_\mu^\nu(-\tilde{k})T_{42}\}, \qquad (25)$$

$$N_h = 1/2 + \sum \sum_j \eta_j T_{jj}. \qquad (26)$$

Here $\sum$ is sum over $k < F$, i.e. over the half of phase volume and

$$T_{ij}(k) = U_{i\lambda} f_\lambda U_{\lambda j}^+, \quad \{\eta_j\} = \{1,1,-1,-1\}_j, \quad j = 1,\ldots,4. \qquad (27)$$

In case of one type of pairing equations (24–26) reduce to standard ones.

Starting point in choosing the parameters was the requirement of small difference of renormalized HF energies $|\varepsilon_d - \varepsilon_p| < t$ for the half filled upper band, i.e. the close energies of configurations with one holes on Cu or one holes on O sites. This gives any reason for application of the band approach through one expects only the qualitative results.

Original set of parameters for presented 2D calculations taken from [9] are (all in eV)

$$t(N_h = 0) = 1.3, \quad Q = 1.2, \quad U_d = 10.0, \quad U_p = 4.0 \qquad (28)$$

together with value $(\varepsilon_d - \varepsilon_p) = 1.2\,\mathrm{eV}$ [9] at $N_h = 0$. Our study is extended also on cases $(\varepsilon_d - \varepsilon_p) < 0$ down to $\varepsilon_d - \varepsilon_p = -1.5t$ since namely these models reveal an explicit region of superconductivity. Other key parameters can be evaluated as $K_d \sim K_p \sim t$, $J \sim t^2/I \ll t$, where $I = |\varepsilon_d^0 + \varepsilon_p^0|/2$. Main calculations are done for

$$K_d \simeq K_p \simeq t, \quad J = 0 \quad \text{or} \quad K_d = 0, \quad K_p \simeq 1.5t, \quad J = 0. \qquad (29)$$

Calculations incorporate also the reduced values $(U_d, U_p) = (8.0, 3.2)\,\mathrm{eV}$ besides values $(10.0, 4.0)\,\mathrm{eV}$.

Solution of HF equation (8) gives approximately linear dependence of the band parameters on doping

$$t(H_h) = t_0(N_h) + t_1 N_h; \quad E_d(N_h) = \varepsilon_d - \varepsilon_p = E_d(0) - qN_h, \qquad (30)$$

with $t_1 \simeq 0.5$, $q = 0.6 \div 2.1\,\mathrm{eV}$ for sets of original parameters (Fig. 1 in [18]). Here $N_h = 5 - n_e$ is the hole number per site associated with doping. Variations of E_d with doping play an important role. For other parameters being equal the holding of E_d constant leads to extending of the AF phase region and disappearing of SC one.

As a first step the phase diagrams have been studied for transitions from normal state (N) to each of states with one separate type of pairing (AF or SC-s or SC-d). Dependence $T_c(N_h)$ of the transition temperature upon doping for each type of pairing have been obtained from equations

$$\mathrm{Det}[D_{\nu\nu'}] = 0, \quad D_{\nu\nu'} = [\delta_{\nu\nu'} - \partial\Delta^\nu/\partial\Delta^{\nu'}], \quad \text{at} \quad \Delta^\nu = 0,$$
$$\text{or} \quad D_{\nu\nu'} = [\delta_{\nu\nu'} - \partial\Gamma^\nu/\partial\Gamma^{\nu'}], \quad \text{at} \quad \Gamma^\nu = 0. \tag{31}$$

Indices numerate all OP with given symmetry of weight functions.

At $E_d = 1.2\,\text{eV} > 0$ the N–SC-s boundary can occur partly outside the AF region only for Coulomb integrals U_d, U_p reduced down to values $5.1, 3.1\,\text{eV}$. But at $E_d < 0$ there is a doping region where the SC–N boundary can be partly outside the AF–N boundary even at large Coulomb integrals U_d, U_p if parameters K_d, K_p of the correlated hoping interaction are sufficiently large. Depending on relation K_d and K_p it may be the superconductivity of s-type (SC-s) if $K_d \simeq K_p \simeq t$ or the SC-d if $K_d \ll K_p \simeq 1.5t$. Figures 1,2 present the examples of "zero" phase diagrams.

If $K_d = K_p = t$ the SC-s phase exists as one can see in Fig. 1 for $E_d = -1.0$, $U_d = 10$, $U_p = 4.0$ or $U_d = 8$, $U_p = 3.2$ (all in eV). The N–SC-d boundary lies entirely inside the AF region if it does exist. Situation inverts at $K_d = 0$, $K_p > 0$. The "zero" N–SC-s boundary disappears at $K_p < 2.1\,\text{eV}$, but there is a large SC-d region after destruction of the AF order. Figure 2 demonstrates this phenomena for $E_d = -t = -1.3\,\text{eV}$, $K_d = 0$ and $K_p = 2.2\,\text{eV}$ (solid curves) or $K_p = 2.0\,\text{eV}$ (dashed curves). Other parameters are from (29).

A reality of "zero" phase diagram or more exactly of the right boundaries of N phase is confirmed by solutions of self-consistent equation (24–26) for OP. Iteration procedure for finding the OP works only at fixed number of particles but not at fixed chemical potential and converges to "pure" solution corresponding to only AF or only SC ordering (but not mixed one) or to zero for N state. Thus the general solution can be obtained by finding both "pure" solutions choosing one of them with minimal energy. The range of doping where the SC solution has lowest energy correlates quantitatively with the part of the "zero" SC–N phase boundary outside the region of AF ordering.

In comparison with s type of SC at $K_d \simeq K_p \simeq t$ the d-type superconductivity at $K_d = 0$, $K_p > t$ seems more plausible (if any) for description of real system for next reason. 1) It corresponds to more realistic relatively small scale of T_c and of energy gap G contrary to SC-s. However for both SC-s and SC-d the MF approach gives the ratio $G/KT_c < 1$ instead of experimental values $2.4 \div 8$ [2] and the BKS prediction 3. 2) Close values of K_d and K_p for such different sites as Cu and O are hardly probable. 3) Strong correlations of electrons on Cu site can suppress the influence of first term in the correlated hopping interaction (3) depending on Cu occupation.

Thus the MF calculations for the upper band in assumption of frozen lower bands give an indication on possibility of superconductivity which is responsible for the local interaction $V_K \sim K_d, K_p$ of correlated hopping

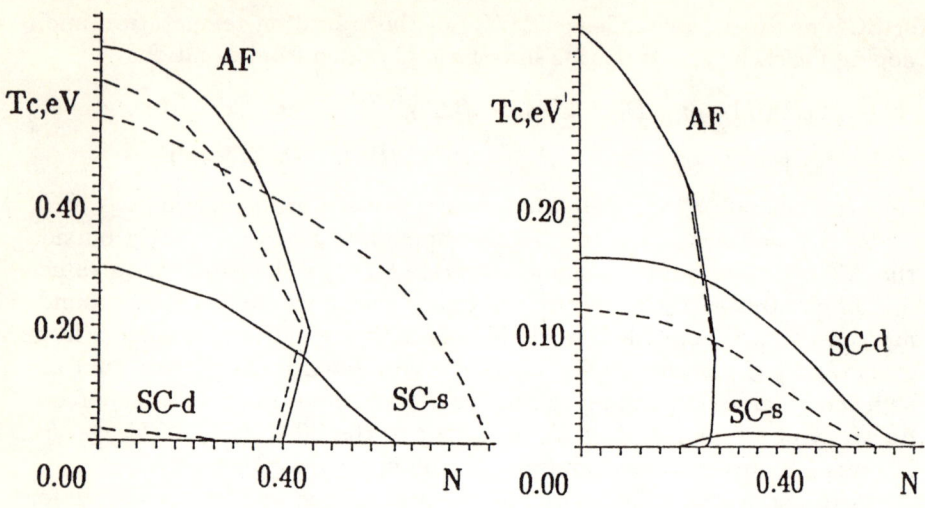

Fig. 1. The "zero" phase curves for AF–N and SC-s – N transition for parameters $E_d(N_h = 0) = -1.0$, $K_d = K_p = t(N_h = 0) = 1.3$, $Q = 1.2$, $J = 0$ and $U_d = 10$, $U_p = 4$ (solid curves) or $U_d = 8$, $U_p = 3.2$ (dashed curves). All values are in [eV]. Soliton for the "zero" N–SC-d boundary exist only for U_d, U_p equal 8,3.2 but lies entirely inside the AF or SC-s region

Fig. 2. The "zero" phase boundaries for the AF–N and SC-d – N transitions for $Q = 1.2$, $J = 0$, $E_d(N_h = 0) = -t(N_h = 0) = -1.3$, $U_d = 10.0$, $U_p = 4.0$, $K_d = 0$, and $K_p = 2.0$ (solid curves) or $K_p = 2.2$ (dashed curves). The SC-s – N boundary inside AF region appears only at $K > 2.1$. All values are in eV

between Cu and O centers. The SC solution appears only at $E_d = \varepsilon_d - \varepsilon_p < 0$ i.e. in case of preferential occupation of O sites by holes and it can be of s-type for $K_d \simeq K_p \simeq t$ or of d-type at $K_d \ll K_p \simeq 1.5t$. The range of doping with the SC follows the region of the AF ordering but solutions with coexistence of AF and SC have not been found.

As it just becomes known for us I.O. Kulik [19] and J.E. Hirsch et al [20–23] have suggested and elaborated the similar ideas about mechanism of SC caused by the the correlated hopping interaction CHI. Many important results have been obtained [20–23] for models described by effective on–site and inter–site parameters. In more detailed picture [19] the SC is caused by CHI connected with direct hopping between p orbits of O sites. Basing on physical argument about increasing of the orbit size during its occupation I.O. Kulik chooses such a sign of the $p - p$ CHI that effective $p - p$ transfer integral increases with occupation of this orbits. In terms of the upper band states spanned on our basis differing from [19] we verify that namely this sign of $p - p$ CHI is needed to provide the possibility of SC. This differs from our $p-d$ CHI for which attractive effect in SC is achieved when occupation of p or d orbits reduces the effective $p-d$ transfer integral. It is very interesting to include the $p - p$ CHI in MF calculations.

References

1. J.G. Bednordz, K.A. Müller: Z. Phys. **B64** 189 (1986)
2. R. Micknas, J. Ranninger, S. Robaszkiewicz: Rev. Mod. Phys. (1990) in press
3. W. Weber: Phys. Rev. Lett. **58** 1371 (1987)
4. V.J. Emery: Phys. Rev. Lett. **58** 2794 (1987)
5. C.M. Varma, S. Schmitt-Rink, E. Abrahams: Sol. St. Commun. **62** 861 (1987)
6. J.E. Hirsh: Phys. Rev. Lett. **59** 228 (1987); **60** 380 (1988)
7. E.B. Stechel, D.R. Jennison: Phys. Rev. **B38** 1634 (1988)
8. J. Zaanen, O. Jepsen, O. Gunnarson, A.T. Paxton, O.K. Andersen, A. Svane: Physica **C153–155** 1636 (1988)
9. M.S. Hybertsen, M. Schluter, N.E. Christensen: Phys. Rev. **B39** 9028 (1988)
10. H. Rushan, C.K. Chew, K.K. Phya, Z.Z. Gan: Phys. Rev. **B39** 11653 (1989)
11. I.I. Mazin: Uspekhi Fiz. Nauk. **158** 155 (1989)
12. J. Friedel: OGLHTCS Preprint **12** 1 (1989)
13. R. Micknas, J. Ranninger, S. Robaszkiewicz, S. Tabor: Phys. Rev. **B37** 9410 (1989)
14. R. Micknas, J. Ranninger, S. Robaszkiewicz: Phys. Rev. **B39** 11653 (1989)
15. Dong–ning Sheng, Chang–de Gong J.: Phys. Condens. Matter **1** (1989)
16. Yu.V. Kopaev: Uspekhi Fiz. Nauk **159** 568 (1989)
17. A.M. Oles, J. Zaanen: Phys. Rev. **B39** 9175 (1989)
18. A.A. Ovchinnikov, M.Ya. Ovchinnikova: in press
19. I.O. Kulik: Superconductivity: Physics Chemistry Technics **2** 175 (1989) [in Russian]
20. J.E. Hirsh: Phys. Lett. **A134** 451 (1989); **136** 163 (1989); Phys. Lett. **A138** 83 (1989)
21. J.E. Hirsh, S. Tang: Phys. Rev. **B40** 2179 (1989)
22. J.E. Hirsh, F. Marsiglio: Phys. Rev. **B39** 11515 (1989)
23. F.M. Marsiglio, J.E. Hirsh: Phys. Rev. **B41** 6435 (1990)

Correlation Pairing and Antiferromagnetic Phase Energy in Low–Dimensional Systems of La–Sr–Cu–O and Y–Ba–Cu–O Metaloxides

I.I. Ukrainskii and E.A. Ponezha

Institute for Theoretical Physics, Metrologicheskaya 14, 252130 Kiev, USSR

1. Introduction

Studies of the ground state of low–dimensional electron systems are now of great interest. This fact is, in particular, a result of synthesis of new high-T_c superconductors (HTSC) [1–3], the HTSC theory is in progress. In accord with a number of papers [4, 5], the htsc can be explained by electron pairing due to electron correlation effects. Correlation electron pairs appear in 1–d [6–8] and 2–d Hubbard models [4, 5].

So, the investigation of the ground state properties of the Hubbard model in one and two dimensions and comparing them with properties of real HTSC La–Sr–Cu–O and Y–Ba–Cu–O are of interest. Studying these properties in the present paper, we use for the many–electron wave function the approach of varying localized geminals (VLG), suitable for description of correlation pairing in 1–d and 2–d systems [5–8]. Our treatment is also a single–band model.

2. The Properties of Half–Filled Conduction Band and Singlet Pairing Energy

We consider the unique band model with the Hamiltonian

$$\widehat{H} = \widehat{T} + \widehat{V} = \sum_{pqp'q'\sigma} t_{pqp'q'\sigma} \widehat{c}_{pq\sigma} \widehat{c}_{p'q'\sigma} + u_0 \sum_{pq} \widehat{c}^+_{pq\uparrow} \widehat{c}_{pq\uparrow} \widehat{c}^+_{pq\downarrow} \widehat{c}_{pq\downarrow} \\ + u_1 \sum_{pq\sigma'\sigma} \widehat{c}^+_{pq\sigma} \widehat{c}_{pq\sigma} c^+_{p\pm1,q\pm1,\sigma'} \widehat{c}_{p\pm1,q\pm1,\sigma'} , \quad (1)$$

where $\widehat{c}^+_{pq\sigma}$ is the electron creating operator, σ is a spin variable and p, q – are the 2–d discrete site coordinates, u_0 and u_1 are the electron–electron repulsion parameters, $t_{pqp'q'}$ is the parameter of conduction band width.

We consider (1) on a simple square lattice, with a being the lattice constant. Let us introduce the operators diagonalizing the kinetic energy in (1)

$$\sum_{pqp'q'\sigma} t_{pqp'q'} \widehat{c}_{pq\sigma} \widehat{c}_{p'q'\sigma} = \sum_{k\sigma} \varepsilon_k \widehat{a}^+_{k\sigma} \widehat{a}_{k\sigma} , \qquad (2)$$

where

$$\widehat{a}_{k\sigma} = N_a^{1/2} \sum_{pq} \exp(i\boldsymbol{k}\boldsymbol{R}_{pq}) \widehat{c}^+_{pq\sigma} ,$$

$$\varepsilon_k = -2t \cos k_x a + \cos k_y a , \qquad (3)$$

$$\boldsymbol{k} = \{k_x, k_y\} \equiv \left\{\frac{n_x}{N_x}, \frac{n_y}{N_y}\right\} \frac{2\pi}{a} ,$$

$$-\frac{1}{2} N_x \le n_x, n_y \le \frac{1}{2} N_y, \quad N_x = N_y = \sqrt{Na} . \qquad (4)$$

We suppose that the electron number Ne is equal to the number of sites Na, i.e. electron density $\rho = Ne/Na = 1$.

We study the energy and the wave function of the ground state for the case of strong and intermediate Hubbard repulsion strength

$$u > \{\max \varepsilon_k - \min \varepsilon_k\} = 4t . \qquad (5)$$

Now we explore the VLG approach [5–8]. The VLG wave function is the best variational wave function for 1–d Hubbard Hamiltonian at any value of u: $-\infty < u < \infty$ [6].

The extension of the VLG in two dimensions was performed by one of us in Refs. [4, 5]. Here we list the main expressions. The ψ_0-function of the ground state of the half–filled conduction band has the form [5]:

$$\widehat{\psi}_0^{(\text{VLG})} = \prod_m (u \widehat{f}^+_{m\uparrow} \widehat{f}_{m\downarrow} + v \widetilde{\widehat{f}}^+_{m\uparrow} \widetilde{\widehat{f}}_{m\downarrow}) |0\rangle = \prod_m \widehat{G}^+_m |0\rangle , \qquad (6)$$

where $u = \cos\varphi$, $v = \sin\varphi$, φ are the variational parameters, Fermi-operators $\widehat{f}^+_{m\sigma}$, $\widetilde{\widehat{f}}^+_{m\sigma}$ correspond to electron states localized near the point $\boldsymbol{R}_m$:

$$\widehat{f}_{m\sigma} = \left(\frac{2}{Na}\right)^{1/2} \sum_k e^{-i\boldsymbol{k}\boldsymbol{R}_m} \widehat{A}^{(1)}_{k\sigma} ,$$

$$\widetilde{\widehat{f}}_{m\sigma} = \left(\frac{2}{Na}\right)^{1/2} \sum_k e^{-i\boldsymbol{k}\boldsymbol{R}_m} \widehat{A}^{(2)}_{k\sigma} , \qquad (7)$$

$$\widehat{A}^{(1)}_{k\sigma} = \widehat{a}_{k\sigma} \cos\theta_k + \widehat{a}_{\widetilde{k}\sigma} i \sin\theta_k ,$$

$$\widehat{A}^{(2)}_{k\sigma} = \widehat{a}_{\widetilde{k}\sigma} \cos\theta_k + \widehat{a}_{k\sigma} i \sin\theta_k , \qquad (8)$$

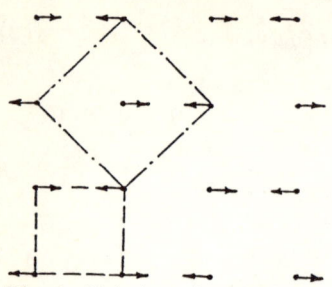

Fig. 1. Unit (- - -) and magnetic (· — · — ·) cells on a quadratic lattice

where $\widetilde{\boldsymbol{k}}$ is the unique vector function of $\boldsymbol{k}$ [5], sums are restricted by the Fermi–level, θ_k are the variational parameters. The $\boldsymbol{R_m}$ vectors form the lattice commensurate with the initial one.

Here we need the concrete lattice and Fermi–surface. We consider the simple square lattice (Fig. 1) with a half-filled conduction band corresponding to the case of ideal nesting [5].

In this case we have for $\widetilde{\boldsymbol{k}}$ [5]

$$\widetilde{\boldsymbol{k}} = \{\widetilde{k}_x, \widetilde{k}_y\} = \{k_x - \frac{\pi}{a}\mathrm{sign}k_x, k_y - \frac{\pi}{a}\mathrm{sign}k_y\} \,. \tag{9}$$

Using the VLG approach we deal with one-electron functions corresponding to $\widehat{f}_{m\sigma}$ operators (7)

$$f_m(\boldsymbol{r}) = f(\boldsymbol{r} - \boldsymbol{R_m}), \quad \widetilde{f}_m(\boldsymbol{r}) = \widetilde{f}(\boldsymbol{r} - \boldsymbol{R_m}) \,. \tag{10}$$

It is of importance that these functions obey orthonormalization conditions

$$(x_m - x_{m'}) \pm (y_m - y_{m'}) = 2al \,, \tag{11}$$

where l is the integer number, x_m and y_m are coordinates of $\boldsymbol{R_m}$ vector. The relations (11) are satisfied when [5]

$$\boldsymbol{R}_{mn} = a\{n + \delta_x, n + \delta_y + 2m\} \,, \tag{12}$$

where m and n are integer and $\delta_{x,y}$ is real. One of possibilities

$$\theta_k = \lambda_x k_x + \lambda_y k_y \tag{13}$$

leads to a number of generated structures [5]

$$\begin{array}{llll} \delta_x = \pm\frac{1}{2}, & \delta_y = 0, & \lambda_x \neq 0, & \lambda_y = 0, \\ \delta_x = 0, & \delta_y = \pm\frac{1}{2}, & \lambda_x = 0, & \lambda_y \neq 0 \,. \end{array} \tag{14}$$

In this case the ? vectors are situated between atoms, i.e. at bonds.

The energy of the ground state (6) per electron pair has the form

$$\varepsilon = \frac{E}{Nt} = -\varepsilon_g + \frac{u}{2} \,, \tag{15}$$

where
$$\varepsilon_g = \sqrt{4t^2 + K^2}, \tag{16}$$
where t and K are the main values of kinetic and interaction energies
$$t = \langle 0|\hat{f}_{m\sigma}\hat{T}\hat{f}^+_{m\sigma}|0\rangle, \tag{17}$$
$$K = \langle 0|\hat{f}_{m\uparrow}\hat{f}^+_{m\downarrow}\hat{V}\hat{f}^+_{m\downarrow}\hat{f}^+_{m\uparrow}|0\rangle. \tag{18}$$
The values of t and K depend on λ-parameter in (13). Thus
$$t^{(2d)}(\lambda) = \frac{16}{\pi^2}\frac{\cos^2\lambda\pi}{(1-4\lambda^2)^2}, \tag{19}$$
$$K^{(2d)}(\lambda) = 2u\sum_n |f_m(\boldsymbol{r}_n)|^4. \tag{20}$$
In 1–d case the expression (16) is also valid, but
$$t^{(1d)}(\lambda) = -\frac{4}{\pi}\frac{\cos\lambda\pi}{1-4\lambda^2}, \tag{21}$$
$$K^{(1d)}(\lambda) = \frac{u}{3}(1 + \frac{1}{2}\sin\lambda\pi). \tag{22}$$
The u-dependence of ground state energy is given in Fig. 2.

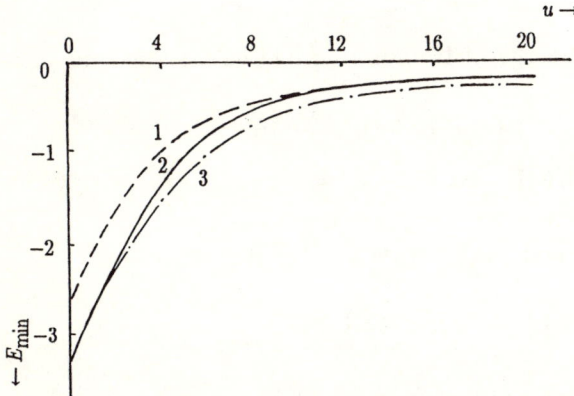

Fig. 2. The ground–state energy of one–dimensional (1) and two–dimensional (2) Hubbard systems in the VLG approximation (6); (3) – the energy of a two–dimensional system in the approximation (23)

3. Antiferromagnet Phase Energy

The VLG wave function (6) is built up from doubly-filled configurations only. This form leads to bosonization of excitations. But the system with Hubbard Hamiltonian has a tendency to destruct electron pairs, resulting in low–lying triplet excitations which are homopolar or noncurrent. The above fact was established in studying 1–d Hubbard systems [9].

These properties simply that for large interaction $u \gtrsim t$ the correlation corrections to the ground state energy can be obtained including the configurations with unpaired electrons into the ground state wave function. In the case of attractive forces they are singlet states and in the case of repulsion–triplet ones. Thus, the VLG wave function now is

$$\widehat{\psi}_{\text{VLG}} = \Pi \widetilde{\widehat{G}}_m |0\rangle , \qquad (6a)$$

where

$$\widetilde{\widehat{G}}_m = [u\widehat{f}^+_{m\uparrow}\widehat{f}^+_{m\downarrow} + v\widetilde{\widehat{f}}^+_{m\uparrow}\widetilde{\widehat{f}}^+_{m\downarrow} + w(\widehat{f}^+_{m\uparrow}\widetilde{\widehat{f}}^+_{m\downarrow} - \widetilde{\widehat{f}}^+_{m\uparrow}\widehat{f}^+_{m\downarrow})] , \qquad (23)$$

and

$$u = \cos\varphi\cos\psi, \quad v = \sin\varphi\cos\psi, \quad w = \frac{1}{\sqrt{2}}\sin\psi . \qquad (24)$$

Varying the ground state energy of Hamiltonian (1) with respect to φ, we obtain

$$\begin{aligned}
E &= -2t(u^2 - v^2) + 2uvK + (u^2 + v^2)K \\
&\quad + \left(\frac{u}{2} - K\right)[1 - 4w^2(u-v)^2 = \\
&= (\sqrt{4t^2 + [k + 2(\frac{u}{2} - K)\sin^2\psi]^2} - K)\cos^2\psi \\
&\quad + \left(\frac{u}{2} - K\right)\left(1 - \frac{1}{2}\sin^2 2\psi\right) ,
\end{aligned} \qquad (25)$$

where notations are the same as in expressions (21).

From the expression (25) we see that the ground state energy is lower when triplets are included in the wave function depending on the values of φ and λ in (16).

Triplet configurations in the ground state (6a) lead to magnetization of sites with antiferromagnet ordering (Fig. 1b)

$$\begin{aligned}
S^z_{m,n} &= (-1)^{n+m} w(1 - 2w^2)^{1/2} 2\sqrt{2} , \\
\widehat{S}^z_{m,n} &= \frac{1}{2}(\widehat{c}^+_{nm\uparrow}\widehat{c}_{nm\uparrow} - \widehat{c}^+_{nm\downarrow}\widehat{c}_{nm\downarrow} .
\end{aligned} \qquad (26)$$

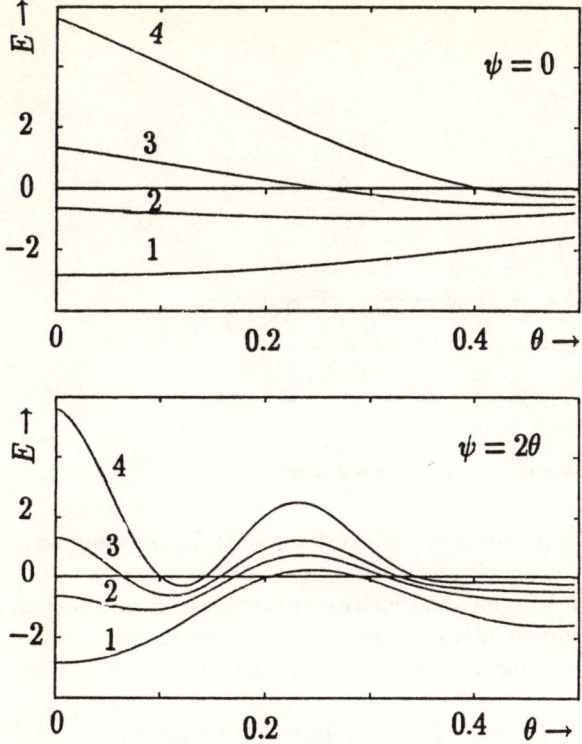

Fig. 3. The energy surface cross-sections along the lines $\psi = 0$, $\psi = 2\theta$ ($\theta = \pi\lambda$) for different u in the expression (1): 1 – $u = 2$, 2 – $u = 6$, 3 – $u = 10$, 4 – $u = 20$

As a result of antiferromagnet spin interaction between Cu–O planes, we obtain the antiferromagnet structure established in experimental study of La_2CuO_4 [10].

Now consider the ground state energy dependence on the parameters of the system. Fig. 3 shows the ψ- and ($\theta = \pi\lambda$)-dependence of the ground state energy. We can see that θ- and ψ-dependencies are characterized by a number of local minima in contradiction to θ-dependence at $\psi = 0$, which is a very smooth curve. Let us consider the structure of energy surface in the presence of antiferromagnetic ordering. The θ- and ψ-dependencies of the ground state energy E is shown in Fig. 3. We can see that θ-dependence is characterized by a number of local extremes in contrast to the θ-dependence of E along $\psi = 0$. The oscillating character of ψ-dependence of E complicates the search of the absolute minimum.

The complicated form of the energy surface is shown in the three-dimensional picture (Figs. 5, 6). Figs. 5 and 6 show the θ- and ψ-dependence of the ground state energy (25) under constraints $\partial E/\partial \varphi = 0$, $\partial^2 E/\partial \varphi^2 = 0$.

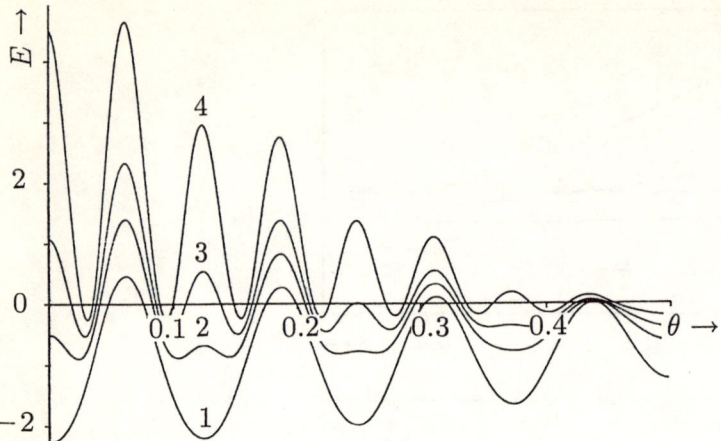

Fig. 4. The energy surface cross section along the line $\psi = 8\theta = 8\pi\lambda$

This means that the energy is minimized with respect to φ for different values of u.

We can see in Figs. 5 and 6 that the energy surface has a number of valleys and ridges. The movement along valleys is equipotential with accuracy of a few percents. But in the normal direction there visible potential barriers.

We can make the next conclusions. The total minimum – i.e. the ground state – is situated always in one of valleys. The total minimum can be easily moved along the valley by small perturbations such as magnetic interactions between planes, small doping. This effect is a result of quasi–degeneration of the states in the valley and indicates a presence of a hidden symmetry in the 2–d Hubbard model.

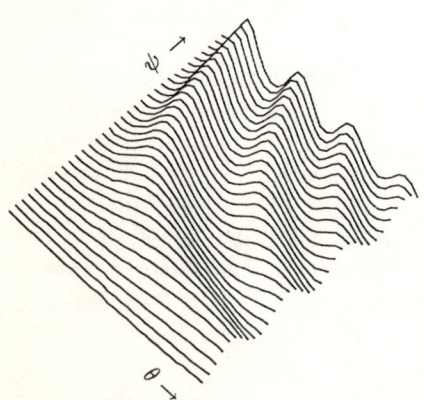

Fig. 5. The general view of an equipotential surface for $u = 2$

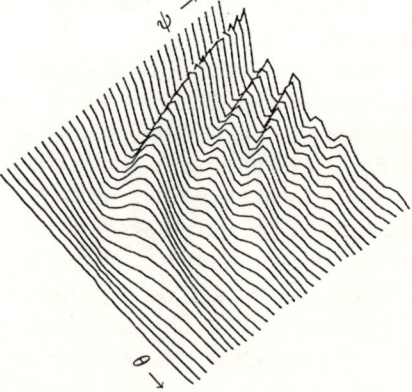

Fig. 6. The general view of an equipotential surface for $u = 10$

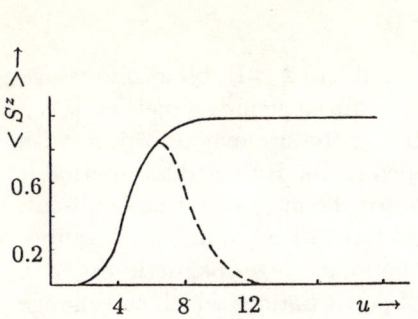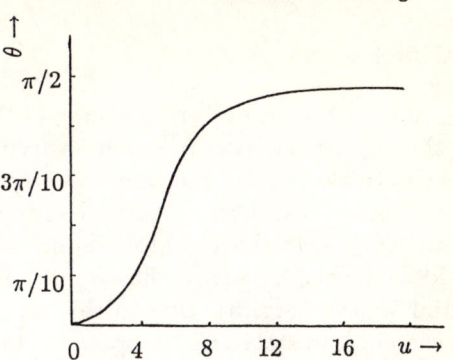

Fig. 7. The magnetization of the node in the structure in Fig. 1 as function of u. The dashed line shows the uncertainty in $\langle S^z \rangle$ due to degeneration of the ground state

Fig. 8. The dependence of the optimum value of $\theta = \pi\lambda$ on u

The position of the absolute minimum – i.e. the deepest valley – depends on the u–parameter of (1). When $u < 4$ the total minimum lies in the first valley near the axis $\psi = 0$. Then, with increase of $u > 4$ it moves to the valley near the line $\psi \approx \pi/4$. When $u > 9$ we have reverse movement, but also there exists quasi–degeneration giving rise to uncertainty of the minimum position $0 < \psi < \pi/4$.

The change of θ- and ψ-parameters is connected with the change of the ground state properties. Thus, the lattice magnetization is proportional to $\sin 2\psi$. So, we obtain the u-dependence of the site magnetization (26), given in Fig. 7.

Comparing our data with the experiments on the site magnetization in $LaCu_2O_4$ [10] and $YBa_2Cu_3O_{6+x}$ [11], we can conclude that in this materials $u = 4.5$.

In 1-d case the total minimum corresponds to $\psi = 0$, i.e. zero magnetization at any value of u. But the energy values in different valleys are nearly the same.

So we can conclude that displacements of total minimum from one valley to another are possible as a result of small perturbations. So, the glass–like behavior of 1-d and 2-d Hubbard systems is possible. Doping of the system and appearance of current carriers leads to the decrease of magnetization as a result of factor $(1 - w^2)$ in the kinetic energy of carriers [5]. On the other hand, the magnetic interactions between Cu–O planes in the form

$$\sum J_{\text{eff}} s_i s_j$$

leads to the stabilizations of the magnetic long-range order even though the value of J_{eff} is smaller by order than its value in plane.

4. Conclusions

We consider the ground state energy of the 1–d and 2–d Hubbard model using the approach of pair function – varying localized geminals method [5–8]. It is shown that in the 2–d Mott semiconductor the site magnetization with antiferromagnet ordering sharply increases when the Hubbard interaction is of order $u \geq 4$. In the 1–d Mott semiconductor the magnetization is absent. Besides, the energy surface has complicated form with a number of equipotential valleys (Fig. 5). Due to this fact the long–range magnetic order in the system is unstable with respect to small perturbations which can change essentially the site magnetization. The examples of such perturbations can be small doping, chain–plain and chain–chain interactions. This instability also can lead to spin–glass behavior of the system.

Comparison of our calculations (Fig. 4) with the experiment for La–Cu–O and Y–Ba–Cu–O [10,11] allows us to conclude that in these systems $u \gtrsim 6.5$.

References

1. J.C. Bednorz, K.A. Muller: J. Phys. **B64** 189 (1986)
2. M.K. Wu, J.B. Ashburn, C.J. Torng et al.: Phys. Rev. Lett. **58** 1574 (1987)
3. J.Z. Sum, D.J. Wedd, M. Natio et al.: Phys. Rev. Lett. **58** 1574 (1987)
4. P.W. Anderson: Science **235** 1196 (1987)
5. I.I. Ukrainskii: Fizika nizkikh temperatur **19** 883 (1987) [Sov. J. Low Temp. Phys. **13** 507 (1987)]
6. I.I. Ukrainskii: Teor. Mat. Fizika **32** 392 (1977) [Sov. physics TMP **32** 392 (1977)]
7. I.I. Ukrainskii: Zh. eksp. teor. fisika **76** 760 (1979) [Sov. physics JETP **49** 381 (1979)]
8. I.I. Ukrainskii: Phys. Stat. Solidi **b 106** 55 (1981)
9. A.A. Ovchinnikov, I.I. Ukrainskii, G.F. Kventsel: Usp. fiz. nauk **108** 81 (1973) [Sov. Phys. Usp. **15** 575 (1973)]
10. D. Vaknin, S.K. Monston, D.C. Jonston, J.M. Newsam, C.R. Satinya, H.E. King: J. Phys. Rev. Lett. **58** 2802 (1987)

Kink Nature of Current Carriers in High-T_c Superconductor–Oxides

I.I. Ukrainskii, M.K. Sheinkman and K.I. Pokhodnia

Institute for Theoretical Physics, Metrologicheskaya 14 252130 Kiev, USSR

1. Introduction

Much interest has been generated in studying the electronic structure and physical properties of cuprate oxides–high-T_c superconductors (HTS) of La–Cu–O and Y–Ba–Cu–O types [1, 2]. Some properties of HTS still remain unclarified. In particular several different mechanisms and model of current carriers have been proposed: band conductivity, polaron model, holon mechanism, kink model and so on.

In the present paper we discuss the theoretical and experimental evidence in favour of a kink nature of current carriers in HTS La–Cu–O, Y–Ba–Cu–O and Bi–Ca–Ti–Cu–O.

The kink model (KM) of HTS was developed by one of us using the localized geminals approach [3–5]. The kink model shares some feature with the RVB model of Anderson [6]. Within both models the superconductivity of HTS is due to electron correlations effects. However, the ground state wave function in the kink model essentially differs from that in the RVB model. Thus, the kink model allows Peierls phonon condensation [93,4], which is of importance for the kink properties.

Earlier the existence of a kink state was assumed in conducting organic polymers which are quasi–one–dimensional systems based on conjugated molecules such as polyacetylene [7]. In section 2 and 3 of this paper we describe the theoretical background of kink states in 2–d systems, employing our previous results [3,4].

2. Kink Properties in High-T_c Oxides

The kink model in its simplest form can be derived from the 2–d Hubbard Hamiltonian

$$\mathcal{H} \sum m,l,\sigma t_l c^+_{m+l,\sigma} c_{m\sigma} + U \sum_m c^+_{m\uparrow} c_{m\uparrow} c^+_{m\downarrow} c_{m\downarrow} , \qquad (1)$$

where $c^+_{m\sigma}$ denotes the electron creation operator at the m-th site with σ-spin. As lattice sites in La_2CuO_4 and $YBa_2Cu_3O_7$ oxides we take an electronic orbital with a $3d(Cu) - 2p(O) - \sigma$-type hybridization. These sites form a simple square lattice. The conduction band width t-parameter in Hamiltonian (1) depends on the distance between the nearest sites. A uniform square lattice gives the ideal nesting in a half–filled conduction band. Using the method of varying localized geminals (VLG) [8–10] for solving the many–electron Schrödinger equation, we can conclude that 2–d electrons form pairs [3,4]. The total wave function takes form of a pair function product

$$\psi_0 = \prod_m G^+_m |0\rangle , \qquad (2)$$

$$G^+_m = (u f^+_{m\uparrow} f^+_{m\downarrow} + v \tilde{f}^+_{m\uparrow} \tilde{f}^+_{m\downarrow}) , \qquad (3)$$

where the Fermi operators $f_{m\sigma}$ and $\tilde{f}_{m\sigma}$ correspond to orthonormalized electronic states similar to bonding and antibonding orbitals in molecules [3, 4, 8–10]. These states are solutions to the Schrödinger equation and are delocalized on the whole 2–d crystal with partial localization near the m site or bond. This partial localization is defined by the expression

$$f_m(\boldsymbol{r}) = \frac{4a^2}{\pi\sqrt{2}} \frac{\sin(\pi/2a)(R_x - R_y)\sin(\pi/2a)(R_x + R_y)}{(R_x^2 - R_y^2)} , \qquad (4)$$

where $\boldsymbol{R} = \boldsymbol{R} - \boldsymbol{r}$.

The delocalization of the f-functions (4) defines an essential difference between the KM [3, 4] and RVB models. The other difference is that one of quasi valent structures in the KM is frozen as a result of the electron–phonon interaction. In the ground state (2), the 2–d system under consideration is the Mott–Peierls semiconductor for any positive values of the Hubbard repulsion parameter in (1) $U > 0$. The forbidden gap width ΔE_g contains both the Mott and Peierls contributions [9, 10]. When $U \gtrsim 4t$, the main contribution to ΔE_g is the Mott (electronic) contribution. This electronic contribution results from the energy need for a destruction of two electron pairs in ground state (2) [10]. This pair destruction process can be described by replacing two geminals in the VLG product (2) with the product of

$$G^+_m G^+_{m+n} \to f^+_{m+n,\sigma} \tilde{f}^+_{m,-\sigma} f^+_{m\uparrow} f^+_{m\downarrow} \qquad (5)$$

with m and n being summed up. The energy of such a state is higher than the ground state energy by the doubled pair binding energy [10] ($2/9 U \leq K \leq 1/2 U$)

$$\Delta E_g \equiv 2\varepsilon_g = 2\sqrt{4t^2 + K^2}, \quad (K \approx 1/2U) . \qquad (6)$$

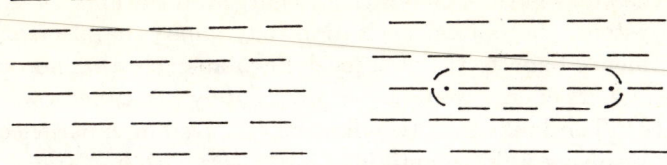

Fig. 1. Graphic representation of the ground state (2). The centers of pair functions are denoted with short lines. The kinks are denoted with points. *Left:* the semiconducting ground state; *right:* a state with one kink pair

An excited state with two kinks has the energy lower than (5). The wave function of a system with one kink pair corresponds to the following substitution in the wave function of the ground state

$$G_m^+ G_{m+n}^+ \prod G_p^+ \to K_m^+ K_{n+m} \prod \widetilde{G}_p^+ , \qquad (7)$$

where the KM kink operator is related to its charge, q [4]

$$K_m^+ = \begin{cases} 1, & (q=1) , \\ c_{m\sigma}^+, & (q=0) , \\ c_{m\uparrow}^+ c_{m\downarrow}^+, & (q=-1) . \end{cases} \qquad (8)$$

The operator $\widetilde{G}_p^+$ in (8) corresponds to pairs with shift (see Fig. 1b, dashed line) [4], in presence of kink, the energy of optical excitation of an electron from the valence band $\hbar\omega_1$ is approximately two times lower than the value (6), because in this case it is enough to break only one pair (see Fig. 2)

$$\hbar\omega_1 \approx \varepsilon_g = \Delta E_g/2 . \qquad (9)$$

Another transition $\hbar\omega_2$ is due to the optical excitation of a neutral kink-antikink pair in the conduction band, i.e. in the state (5).

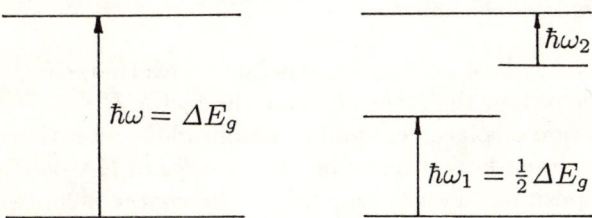

Fig. 2. The optical excitation energies in the semiconducting phase (*left*) and in the state with kinks (*right*)

The occurrence of a kink states in the ground state of the system is connected with doping – deviation from a complete filling of the conduction band [3,4]. The kink are spinless and can carry the charge ±1. In La_2CuO_4

and $YBa_2Cu_3O_{7-y}$ the kinks have to be positively charged. If the kink–holes are present in the system, the optical excitation may cause the electron transition from the valence band to the kink level. Such an excitation needs one pair to be broken i.e. its energy would be approximately two times lower than the gap ΔE_g (6). The kinks in a two-dimensional system experience a confinement [3–5], resulting in kink–antikink (KA) pairs with a charge of $2e$. These KA pairs form a narrow conduction band of t^2/U in width. The radius of KA pair at $U > t$ is of the order $\pi a(U/t)^{1/3}$.

Charged kink disturbs the valence of Cu and O and consequently it will be followed by a local deformation of Cu–O bonds and charged in Cu–O vibration frequencies. In the zone state, as a result, the kink will be enveloped with a fur–coat of a local deformation.

Thus, if the kink model is really true for HTS, the presence of current carriers–kink and KA pairs –in a semiconducting phase inevitably brings new absorption bands into existence, with energies lower than the forbidden gap and leads to some restructuring of the phonon spectrum. Such experiments have been carried out for $YBa_2Cu_3O_{7-y}$ [11,13,15], La_2CuO_4 [12] and $Tl_2Ba_2Cd_{1-x}CuO_8$ [14].

3. Kink Photo–Generation

By exciting the interband transition with the light $\hbar\omega > \Delta E_g$ in a semiconducting phase, one can obtain a certain electron–hole (e–h) pair concentration, n_p which depends upon the density of pumping. Electron–hole pair will relax rapidly forming rather stable kink–antikink pairs. The relaxation of a (e–h) or a (h–h) pair into KA pair will occur in the period of the order $t_r \approx \omega_{ph}^{-1}$ where ω_{ph} is the characteristic frequency of CuO vibration affected by the carrier presence (300–500 cm^{-1} in 2-1-4 and 1:2:3 systems, i.e. $t_r \approx 10^{-13}$ sec.). The life time of a neutral KA pair is much longer than t_r due to the Franck–Kondon factor F. The same picture is valid for polyacetylene [7].

The kink–antikink pair may be considered as a polaron with charge $0, \pm 2$ and mass $M_p > 2m_e$. If we obtain the ratio $M_p/2m_e$ the factor $F \sim e^{-r/r_0}$ can be estimated (here r is ion displacement and r_0-amplitude of vibration). The F factor weakens the direct kink absorption. The charge of KA pair is predetermined by its composition. Two holes produce the charge $+2e$, two electrons $-2e$, e–p pair has no charge.

The charged KA pairs result only from doping. while during the optical excitations predominantly neutral KA pair occur, through the charged KA pairs (bipolarons) may occur at very strong pumping as well. So, at sufficient pumping densities a rather high KA pair concentration can be achieved, so that it will be possible to observe experimentally optical transitions connected with the kink formation (Fig. 2b).

Now, we consider some properties of KA pairs which occur during photogeneration. The kink state is spinless and, consequently, it does not contribute to magnetic susceptibility. As was mentioned above, during photogeneration only neutral KA pairs occur because they are formed from e–p pairs. At the same time a slight amount of KA pairs with the charge $2e$ can be produced.

4. Direct Experimental Evidence of Kink Existence

The most reliable data confirming the existence of non–linear excitations (kinks) in conducting organic polymers were obtained by the method of photo–induced absorption (PIA). This method was used for detection of kinks in polyacetylene (for example, [7] and references therein). This method, practically unmodified, was used in several laboratories to investigate $YBa_2Cu_3O_{7-y}$ [11, 13, 15], La_2CuO_{4-y} [12], $Ti_2Ba_2Ca_{1-x}CuO_8$ [14].

The essence of PIA experiments is the following. Exciting the electron–hole pairs in the semiconducting phase of HTS material by a laser with the energy $\hbar\omega > \Delta E_g$, simultaneously tested the absorption in the region $\hbar\omega < \Delta E_g$. The relative charge of absorption coefficient is being measured which is due to non–equilibrium e–p pairs concentration, $\Delta T/T$.

The article presented gives a new interpretation of PIA experiment for all investigated HTS materials ar a result of kink nature of all photoinduced transitions. So, we consider the current carriers in these materials to be kinks (KA pairs).

As we mention above, the general e–p pairs in the period $t_r \approx 10^{-13}$ sec self–localize and transform into neutral KA pairs. Its energetic level lies at the energy of $\hbar\omega_2$ lower than the conductivity band edge.

The pairs with the charge $+2e$, which also appear (in small amount) in the process of photogeneration have energy $\hbar\omega_1 = 1/2\Delta E_g$. All types of PIA transitions in HTS materials are presented in Fig. 3.

It is evident that in PIA spectra of all these materials the transition $\hbar\omega_2$ predicted by our kink model is observed. Its energy is 0.5; 0.13 and 0.095 eV for La^-, Y^-, Ti^- materials, respectively. Besides, for the two first substances the transition with energy $\hbar\omega_1$ is also observed (1.4 eV and 1.0 eV, respectively).

The kink nature of the states in the forbidden gap and its pair character are confirmed by the nonlinear dependence of the PIA spectra intensity on the pumping power $\Delta T/T \sim I^n$, where $n = 0.5$ [13]. One absorbed quantum gives a pair of kinks.

Another evidence of the kink nature of carriers is the resonance character of 505 cm^{-1} band in the Raman spectrum of thetragonal phase, obtained by IR laser excitation ($\hbar\omega = 1.16$ eV) [15]. In the framework of our model this excitation energy is close to the absorption energy of a double–charged KA pair and thus the resonance conditions have to be satisfied. The fact that

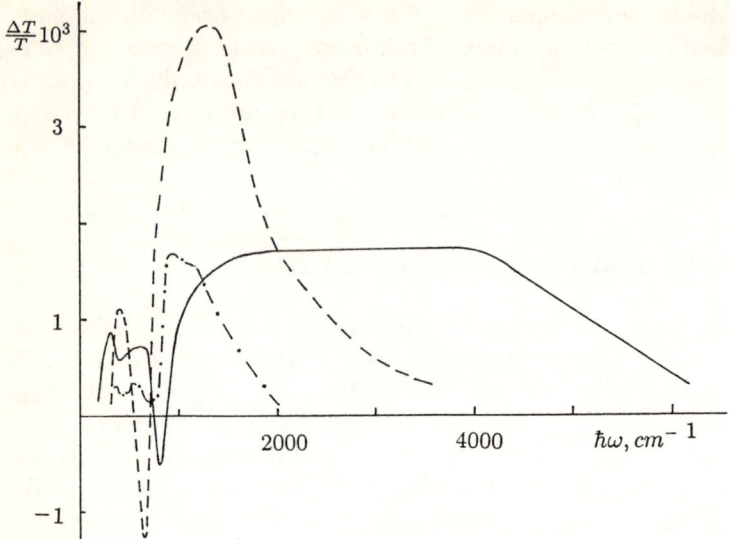

Fig. 3. Photo-induced absorption spectra of La_2CUO_4 (—), $YBa_2Cu_3O_{7-y}$ (- - -) and $Tl_2Ba_2Ca_{1-x}Gd_xCu_2O_8$ (· — · — ·) [11–15]

the frequency of this mode – Ag stretching of Cu(1)–O(4) – is close to the frequency of this mode $YBa_2Cu_3O_7$ shows that the thetragonal lattice has local orthorombic distortions near a KA pair. Besides, just this mode has the highest intensity in PIA spectra [12]. With an increase in $(KA)^{+2e}$ pairs (e.g. due to doping) the system transforms into a metallic–superconducting, at $T < T_c$, state. If the carriers in a superconducting state are KA pairs only, the bands in Fig. 3 characteristic of them, have to be present in the optical spectrum of this phase also, possibly, slightly shifted. They are observed in a spectra of La–Sr–Cu–O (1.0 and 0.3 eV) and in $Y - Ba_2 - Cu_3 - O_7$ (1.4 eV and 0.4 eV). The shifting and widening of these lines can be connected with the filling of the KA pair band and structural changes in superconducting state. As we already mentioned, the doping in our model causes no increase in paramagnetic susceptibility which was also in photo–induced excitations also speaks in favour of its kink nature.

5. Conclusions

The data presented above allow us to make the following conclusions.

In dielectric phases of $La_2CuP_4^-$ and $YBa_2Cu_3O_6$ – HTS the optically induced transitions correspond to energies two times lower than the energies of interband transitions. We can explain this transition, using the kink model [3, 4], as a transition from the valence band to the kink level (Fig. 2b). The analogy with polyacetylene [7] also confirms our conclusion. So, the optical excitation of HTS generates kink states. Then we can conclude that current carriers in HTS are of kink types. More specifically, the carriers are charged kink–antikink pairs due to kink confinement in 2–d systems [3, 4]. The appearance of phonon lines with $\hbar\omega \approx 505\,\mathrm{cm}^{-1}$ is connected with the Peierls phonon mode coupled with kink states [3, 4].

The kink–type properties have also been observed by the other authors in La_2CuO_4 [12] and $T\,L\,Ba\,Gd\,Cu\,O$ [14], $Y\,Ba_2Cu_3O_7$ [11, 13]. The photo-excited e–p pairs in $Y\,Ba_2Cu_3O_7$ [15] are, from our point of view, also neutral kink–antikink pairs.

The kink nature of current carriers in HTS suggests a kink–mechanism of superconductivity. According to this model [3,4], a charged kink is paired with a charged antikink as a result of confinement. Then, the Bose-condensation of KA pairs results in superconductivity.

In contrast, the Anderson RVB model assumes the presence of neutral kinks (spinons) in the ground dielectric state. So, the RVB model cannot explain the photoinduced absorption.

The BCS–type models are not associated with kinks and cannot be useful for a description of the photoinduced transitions in Fig. 2b.

The polaron model can be used for treating the lower transition ($\hbar\omega_2 \leq 0.2\,\mathrm{eV}$) to which polaron effects can be attributed. But, new electronic transitions with energies two $\hbar\omega_1 \approx 1\,\mathrm{eV}$ do not appear in this model.

Thus, the present explanation of photoinduced spectra of HTS on the basis of the kink model seems to be the only possible one.

References

1. J.C. Bednorz, K.A. Muller: J. Phys. **89**(1986)
2. M.K. Wu, J.B. Ashburn, C.J. Torng et al.: Phys. Rev. Lett. **58** 908 (1987)
3. I.I. Ukrainskii: Fiz. nizkich temp. **13** 883 (1987) [Sov. J. Low Temp. Phys. **13** 507 (1987)]
4. I.I. Ukrainskii: Preprint Inst. Teor. Phys. ITP **87** 73 R Kiev (1987)
5. I.I. Ukrainskii: Ac. Sci. USSR, Moscow **248** (1987)
6. P.W. Anderson: Science **235** 1196 (1987)
7. L. Rothberg, T.M. Jedlu, S. Etemad, G.L. Baker: Phys. Rev. **B 30** 7529 (1987)
8. I.I. Ukrainskii: Teor. Mat. Fiz. **32** 392 (1977) [Sov. Phys. TMP **32** 392 (1977)]
9. I.I. Ukrainskii: Zh. eksp. teor. fiz. **76** 760 (1979) [Sov. Phys. JETP **49** 381 (1979)]
10. I.I. Ukrainskii: Phys. Stat. Solidi **b 106** 55 (1981)
11. C.Taliani, R. Zamboni, G. Ruani, A.J. Pal, F.C. Matacone, Z. Vardeny, X. Wea: Proceed. Conf. on High T_c Superconductivity, Rome (1988)
12. Y.H. Kim, A.J. Heeger, L. Acedo, G. Stucky, F. Wudl: Phys. Rev. **B 30** 7252 (1987)
13. C.Taliani, R. Zamboni, G. Ruani, F.C. Matacotta, K.I. Pokhodnia: Sol. State Comm. **66** 487 (1988)
14. C.M. Foster, A.J. Heeger, G. Stucky, N. Herron: Phys. Rev. **B** (1989)
15. B. Zamboni, G. Ruani, A.J. Pal, C. Taliani: Solid State Comm. (1989)

Anomaly Index and Induced Charge on a Noncompact Surface in an External Magnetic Field

Yu. A. Sitenko

Institute for Theoretical Physics, Metrologicheskaya 14, 252 130 Kiev, USSR

The global geometric characteristics influencing fermion charge fractionization on a noncompact space are discussed. In addition to the known characteristics which is related to the curvature tensor (the space part of the total anomaly) we introduce the characteristics which is related to the curvature scalar. The induced vacuum charge on a noncompact simply–connected surface in external magnetic field is found to depend on the latter characteristics and the total magnetic flux.

The inducing of the vacuum quantum numbers by external fields of nontrivial topology generated much interest in the last decades. Since the appearance of the work of Jackiw and Rebbi [1] it has been realized that the quantum states of fermions in the solitonic background can be characterized by the noninteger values of the fermion number (or charge in the units of coupling with the appropriate gauge vector field). This phenomenon, which received the name of the fermion charge fractionization, has observable manifestations in various areas of condensed–matter and particle physics (see the reviews [2, 3]). In the present report we would like to draw attention to the effect of the geometry of space on fermion charge fractionization. The operator of the fermion field which is second–quantized in the background of classical static fields can be represented in the form

$$\Psi(x,t) = \sum_E [\exp(-iEt)\Psi_E^{(+)}(x)b_E + \exp(iEt)\Psi_E^{(-)}(x)d_E^{(+)}] , \qquad (1)$$

where the summation sign implies the summation over the discrete part and the integration over the continuous part of the energy spectrum, $b_E^{(+)}$ and b_E (d_E^+ and d_E) are the fermion (antifermion) creation and annihilation operators satisfying the anticommutation relations, $\Psi_E^{(+)}$ and $\Psi_E^{(-)}$ are the positive– and negative–energy solutions to the equation

$$H\Psi_E(x) = E\Psi_E(x) . \qquad (2)$$

Here H is the Dirac Hamiltonian which in the presence of the external vector field $V_\mu(x)$ in the d-dimensional Riemannian space with the metric $g_{\mu\nu}(x)$ takes the form

$$H = \frac{1}{i}\alpha^\mu(x)\left[\partial_\mu + \frac{i}{2}\omega_\mu(x) - ieV_\mu(x)\right] + \beta m, \tag{3}$$

where

$$[\alpha^\mu(x), \alpha^\nu(x)]_+ = 2g^{\mu\nu}(x)I, [\alpha^\mu(x), \beta]_+ = 0, \quad \beta^2 = I, \quad \text{tr}\,\alpha^\mu(x) = \text{tr}\,\beta = 0$$

and the spin connection is defined as

$$\omega_\mu(x) = \frac{1}{2i}\alpha^\nu(x)[\partial_\mu \alpha_\nu(x) - \Gamma^\rho_{\mu\nu}(x)\alpha_\rho(x)],$$

$$\Gamma^\rho_{\mu\nu}(x) = \frac{1}{2}g^{\rho\lambda}(x)[\partial_\mu g_{\lambda\nu}(x) + \partial_\nu g_{\lambda\mu}(x) - \partial_\lambda g_{\mu\nu}(x)]. \tag{4}$$

The conventional definition of the fermion charge operator yields

$$Q = \frac{1}{2}\int [\Psi^+(x,t), \Psi(x,t)]_- \sqrt{g(x)}d^d x = \sum_E (b_E^+ b_E - d_E^+ d_E) - \frac{1}{2}\eta_H, \tag{5}$$

where $g(x) = \det g_{\mu\nu}(x)$ and the spectral asymmetry of the operator H is defined as

$$\eta_H = \sum_E \text{sgn}(E), \quad \text{sgn}(u) = \begin{cases} 1, & u > 0 \\ -1, & n < 0 \end{cases}. \tag{6}$$

To regularize the divergent sum (integral) in (6) one should introduce a factor $|E|^{-s}$ or $\exp(-tE^2)$ in it and take the limit $s \to 0$ or $t \to 0$ after summing (integrating) over E. The vacuum expectation value of the charge operator is determined by its c-number piece

$$\langle \text{vac}|Q|\text{vac}\rangle = -\frac{1}{2}\eta_H, \tag{7}$$

where

$$b|\text{vac}\rangle = d|\text{vac}\rangle = 0.$$

If the charge of the vacuum state is noninteger, then the same can take place for other (excited) states.

The essential contribution to the spectral asymmetry η_H is given by zero modes of the operator $H - \beta m$. The latter takes the form

$$H - \beta m = \begin{pmatrix} 0 & D^+ \\ D & 0 \end{pmatrix}, \tag{8}$$

in the representation where the matrix β is diagonal

$$\beta = \begin{pmatrix} I & 0 \\ 0 & -I \end{pmatrix}.$$

One can define the quantity

$$\text{Index } D = \text{Dim Ker } D - \text{Dim Ker } D^+ , \tag{9}$$

where Dim Ker D (Dim Ker D^+) is the number of the properly normalized zero modes of the operator $D(D^+)$. Mathematical theorems [4,5] (see the review [6]) relate the value of the index (9) to the topological characteristics of the background manifold – the total anomaly

$$\text{Index } D = A_d . \tag{10}$$

Then a straightforward analysis yields (see, for example, ref. [7])

$$\langle \text{vac}|Q|\text{vac}\rangle = -\frac{1}{2}\text{sgn}(m)\text{Index } D , \tag{11}$$

or

$$\langle \text{vac}|Q|\text{vac}\rangle = -\frac{1}{2}\text{sgn}(m)A_d . \tag{12}$$

Strictly speaking, the relations (10–12) are valid in the case of compact manifolds only. In the case of a closed compact manifold the total anomaly can be represented as the integral over the volume of the base space

$$A_d = \int a_d(x)\sqrt{g(x)}d^d x , \tag{13}$$

where a_d is the Pontryagin–Chern density. The latter vanishes for odd-dimen- sional spaces and is nonvanishing for even–dimensional spaces, for example,

$$a_2 = \frac{e}{4\pi}\varepsilon^{\mu\nu}F_{\mu\nu} , \tag{14}$$

$$a_4 = \frac{e^2}{32\pi^2}\varepsilon^{\mu\nu\rho\sigma}F_{\mu\nu}F_{\rho\sigma} - \frac{1}{768\pi^2}\varepsilon^{\mu\nu\rho\sigma}R_{\mu\nu\delta\omega}R^{\delta\omega}_{\rho\sigma} , \tag{15}$$

where $\varepsilon^{\mu\nu\rho\sigma}$ is the totally antisymmetric tensor,

$$F_{\mu\nu} = \partial_\mu V_\nu - \partial_\nu V_\mu , \tag{16}$$

is the gauge field strength and

$$R^\mu_{\nu\rho\sigma} = \partial_\sigma \Gamma^\mu_{\nu\rho} - \partial_\rho \Gamma^\mu_{\nu\sigma} + \Gamma^\mu_{\delta\sigma}\Gamma^\delta_{\nu\rho} - \Gamma^\mu_{\delta\rho}\Gamma^\delta_{\nu\sigma} , \tag{17}$$

is the curvature tensor. Note that the integral in (13) can take integer values only.

In the case of noncompact manifolds the relations of the type of (10) are lacking. The reason is that zero modes of the operators D and D^+ are not isolated from the continuum in this case. Therefore, in addition to square–integrable zero modes corresponding to bound states the non–square–integrable zero modes corresponding to unbound resonances appear. Thus, threshold effects spoil the Fredholm property of the operator (8), breaking the line of arguments of refs. [4,5]. On the other hand the total anomaly (13) in this case can take integer and noninteger values as well.

Therefore, one can suggest that instead of (11) the relation (12) is valid in the case of a noncompact space (see the corresponding arguments in ref. [8]). Then for both noncompact and closed compact spaces of dimension $d \geq 4$ the induced vacuum charge depends on the spatial geometry. The appropriate global geometric characteristics is the space part of the total anomaly, being related in the case of $d = 4$ to the integral of the quadratic combination of the curvature tensors (15).

However, we find another global geometric characteristics which influences the induced vacuum charge. This characteristics is related to the integral of the curvature scalar

$$R = g^{\mu\nu} R^{\rho}_{\nu\rho\mu} \ . \tag{18}$$

Due to this characteristics, fermion charge fractionization on noncompact manifolds becomes dependent on the spatial geometry even in the case of the space dimension $d = 2$.

To elucidate a loophole in the arguments of ref. [8], let us consider the equation for the divergence of the axial–vector current [9,10]

$$ig^{-1/2} \partial_\mu g^{1/2} J^\mu_{d+1}(x, M) = 2a_d(x) + 2iM J_{d+1}(x, M) \ . \tag{19}$$

Usually, as in ref. [8], the integral of the left–hand side of (19) over a noncompact space is neglected. Then the integration of the right–hand side of (19) over a noncompact space and the parameter M yields (12). However, we find that the contribution of the previously neglected term can be nonvanishing, depending on the geometry of space. The latter circumstance provides us with the global geometric characteristics differing from the space part of the total anomaly.

In the following we confine ourselves to noncompact spaces of dimension $d = 2$. In this case the total anomaly is independent of the spatial geometry, being related to the total flux,

$$A_2 = e\Phi \ , \qquad \Phi = \frac{1}{2\pi} \int B(x)\sqrt{g(x)} d^2 x \ , \tag{20}$$

of the magnetic field strength,

$$B(x) = \frac{2\pi}{e} a_2 \ , \tag{21}$$

which is taken in three–dimensional space to be orthogonal to the two–dimensional space (surface) under consideration. Let us introduce the total integrated curvature

$$\Phi_K = \frac{1}{2\pi} \int K(x)\sqrt{g(x)} d^2 x \ , \tag{22}$$

where K is the Gauss curvature

$$\beta K(x) = \varepsilon^{\mu\nu} \partial_\mu \omega_\nu = \frac{1}{2}\beta R .\tag{23}$$

The index in the space of square–integrable zero modes on a noncompact simply–connected surface is given by the expressions [11]

$\Phi_K > 0:$ Index$D =$

$$\begin{cases} \text{integ}_+(e\Phi + \frac{1}{2}\Phi_K) - 1 , & e\Phi \geq \frac{1}{2}\Phi_K , \\ \text{integ}_+(e\Phi + \frac{1}{2}\Phi_K) + \text{integ}_-(e\Phi - \frac{1}{2}\Phi_K) , & -\frac{1}{2}\Phi_K < e\Phi < \frac{1}{2}\Phi_K , \\ \text{integ}_-(e\Phi - \frac{1}{2}\Phi_K) + 1 , & e\Phi \leq -\frac{1}{2}\Phi_K , \end{cases}\tag{24a}$$

$\Phi_K \leq 0:$ Index$D =$

$$\begin{cases} \text{integ}_+(e\Phi + \frac{1}{2}\Phi_K) - 1 , & e\Phi > -\frac{1}{2}\Phi_K , \\ 0 , & \frac{1}{2}\Phi_K \leq e\Phi \leq -\frac{1}{2}\Phi_K , \\ \text{integ}_-(e\Phi - \frac{1}{2}\Phi_K) + 1 , & e\Phi < \frac{1}{2}\Phi_K , \end{cases}\tag{24b}$$

here integ$_+(u)$(integ$_-(u)$) is the upper (lower) nearest to u integer number. The operator $H^2 - m^2$ is shown to be the Hamiltonian of the supersymmetric quantum mechanics with one fermionic (spin) and two bosonic (coordinates on the surface) degrees of freedom [12]. Since the nonvanishing of the index indicates that the supersymmetry is exact, the answer to the question, whether supersymmetry is spontaneously broken or not, depends on the values of the flux Ψ and the integrated curvature Φ_K as well.

The latter global characteristics influence also fermion charge fractionization. The vacuum charge which is induced on a noncompact simply–connected surface takes the form

$$\langle \text{vac}|Q|\text{vac}\rangle = -\frac{1}{2}\text{sgn}(m)e\Phi - \frac{1}{2}\xi\left[m, \text{fract}\left(e\Phi + \frac{1}{2}\right), \Phi_K\right] ,\tag{25}$$

where

$$\xi(m, u, \Phi_K) = \begin{cases} 0 , & \Phi_K < 1 , \\ \frac{1}{\pi}\arctan[\coth(\pi Rm)\tan(\pi u)] - \text{sgn}(m)u , & \Phi_K = 1 , \\ \text{sgn}(m)\left[\frac{1}{2}\text{sgn}_0(u) - u\right] , & \Phi_K > 1 , \end{cases}\tag{26}$$

$$\text{fract}(u) = u - \text{integ}(u) ,$$

$$\text{sgn}_0(u) = \begin{cases} \text{sgn}(u) , & u \neq 0 , \\ 0 , & u = 0 , \end{cases}$$

$$-\frac{1}{2} < \text{fract}(u) < \frac{1}{2} ,$$

integ(u) is the nearest to u integer number; incidentally one has

$$\text{integ}_\pm(u) = \text{integ}(u) + \frac{1}{2}\text{sgn}_\mp[\text{fract}(u)] \pm \frac{1}{2} , \quad \text{sgn}_\pm(u) = \begin{cases} \text{sgn}(u) , & u \neq 0 , \\ \pm 1 , & u = 0 . \end{cases}$$

The parameter R in (26) is the intrinsic length scale parameter of a surface with $\Phi_K = 1$; in general, this parameter makes the long distance asymptotes of the surface metric in conformal coordinates to be dimensionless.

One can conclude from (25, 26) that the result of ref. [13] holds not only for a plane ($\Phi_K = 0$) but for surfaces with $\Phi_K < 1$ as well. In particular the induced vacuum charge on a plane coincides with that on a hyperbolic paraboloid (saddle–like surface, $\Phi_K = -1$) and differs from that on an elliptic paraboloid ($\Phi_K = 1$), the latter differing also from that on a punctured sphere ($\Phi_K = 2$).

It should be noted that the present consideration can be generalized to arbitrary noncompact Riemannian surfaces.

References

1. R. Jackiw, C. Rebbi. Phys. Rew. **D13** 3398 (1974)
2. A.J. Niemi, G.W. Semenoff. Phys. Rep. **135** 99 (1986)
3. I.V. Krive, A.S. Rozhavskii. Sov. Phys. Uspekhi **30** 370 (1987)
4. M.F. Atiyah, I.M. Singer. Ann. Math. **87** 484 (1968)
5. M.F. Atiyah, V.K. Patodi, I.M. Singer. Math. Proc. Cambridge Philos. Soc. **77** 43 (1975); **78** 405 (1975); **79** 71 (1976)
6. T. Eguchi, P.B. Gilkey, A.J. Hanson. **66** 213 (1980)
7. Yu.A. Sitenko. Sov. J. Nucl. Phys. **47** 184 (1988)
8. M.B. Paranjape, G.W. Semenoff. Phys. Rev. **31** 1324 (1985)
9. S. Adler. Phys. Rev. **177** 2426 (1969)
10. J.S. Bell, R. Jackiw. Nuovo Cim. **60A** 47 (1969)
11. Yu.A. Sitenko. Sov. J. Nucl. Phys. **50** 571 (1989)
12. Yu.A. Sitenko. Problems of Modern Quantum Field Theory (Invit. Lect. of the Spring School in Alushta, USSR, 1989) Eds. A.A. Belavin, A.U. Klimyk, A.B. Zamolodchikov (Springer, Berlin, Heidelberg 1989)
13. A.J. Niemi, G.W. Semenoff. Phys. Rev. Lett. **51** 2077 (1983)

About the Influence of Uniaxial Pressure on the Twin Structure in the 1–2–3 System

V.S. Nikolayev

Institute for Theoretical Physics, Metrologicheskaya 14, 252130, Kiev, USSR

In the present communication we make an attempt to study the influence of the uniaxial stress on the twin structure in the superconducting nonstoichiometric compounds 1–2–3 (a representative of this class that investigated most often is $YBa_2Cu_3O_{7-\delta}$). We also want to make clear the role of lattice deformation that accompanies the twinning.

We proceed our study in the frames of a model of twinning suggested in [1] and used to describe equilibrium twins in the system with $\delta \ll 1$.

Consider briefly the main topics of the model. It is supposed that the major interactions affecting the twin's formation, take place just in the basis plane (the plane of chains) of the 1–2–3 compound. First of all it is the interaction between the ions Cu(1) and O. It is taken into account as a potential for oxygen ions, formed by the rigid Cu lattice (Fig. 1) with the period a_{Cu}:

$$W = \sum_i V\left[1 - \cos\left(\frac{2\pi}{a_{Cu}}x_1^i\right)\cos\left(\frac{2\pi}{a_{Cu}}x_2^i\right)\right]. \tag{1}$$

The second – "concurrent" – interaction is an O–O potential, chosen in a harmonical form which corresponds to the energy of the oxygen ion coordinates $r = (x_1^i, x_2^i)$ deviations from the equilibrium positions (R_i):

$$U = \sum_{i,j}\frac{1}{2}K_{i,j}|r_i - R_i - r_j + R_j|^2, \tag{2}$$

where $K_{i,j}$ are the force constants. R_i corresponds to the coordinates of the sites in the square lattice with the period a_0 which is determined from the condition of equality of areas, occupied by the lattices O and Cu:

$$a_0 = a_{Cu}/\sqrt{1-\delta}. \tag{3}$$

The summation in (1-2) is carried out over all ions O. But when $\delta \ll 1$ we can sum over all oxygen sites in Cu lattice, neglecting the presence of vacancies. We then introduce the displacement $u(n)$ of the ion O with the number n from the n-th oxygen site in the Cu lattice. By going over to

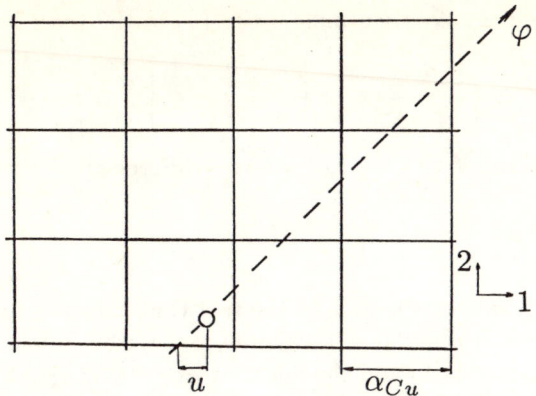

Fig. 1. The structure of the basis plane of a 1–2–3 compound. The Cu-sites are on the lines cross-section, the O-sites are in the middle of the side of the squares. The dotted line shows the direction of O-ions (the circle) displacement from the correspondent position

the continuum ($p = n \cdot a_{Cu}$) and one dimensional (in the direction (1,1)) approximation we obtain

$$H = W + U = \frac{1}{L} \int_0^L d\phi \left\{ 2K \left[\left(\frac{du}{d\phi}\right)^2 - \frac{a_0 - a_{Cu}}{a_{Cu}} \sqrt{2} \frac{du}{d\phi} \right] + V \sin^2 \left(\frac{2\pi u}{a_{Cu}} \right) \right\} \quad (4)$$

where $\phi = (p_1 + p_2)/\sqrt{2}$; $u \equiv u_2 = u_1$; $K \equiv K_{i,i+1}$; $L \to \infty$.

The model described is distinguished from the classical (ferroelastic) one, suggested in [2-4] and based on the standard Ginzburg-Landau functional with the order parameter, proportional to the deformation. The classical model cannot provide for the minima of the volume energy of the periodic structure, without accounting the surface energy [3]. Thus, if we think that the classical model is good, the strong dependence of the twin's period on the dimensions of the monocrystals should be observed. But seemingly such a dependence is absent in the experiment. Besides that, the model [2-4] takes no account of the influence of 1–2–3 compounds nonstoichiometry to the twinning.

The model [1] is deprived of these faults. But in this model the deformation during twinning is not taken into account, meanwhile the majority of the experimental data, obtained in connection with twinning, concerned with the twin deformations, that are observed by optical or optical electronic methods. In this way the authors of Ref.[5] examine the detwinning of 1–2–3 crystal under uniaxial pressure and the authors of [6] study the reaction of the crystal on the point influence.

We introduce into the model [1] the order parameter ε, that means the rhombic deformation (square into rectangle) analogous to that in [2-4]:

$$\varepsilon = (\varepsilon_{11} - \varepsilon_{22})/\sqrt{2}, \quad (5)$$

where ε_{ij} is the deformation tensor. We then must write (4) in the form:

$$H = H|_{\varepsilon=0} + \frac{1}{L}\int_0^L d\phi \left[\frac{Q\varepsilon^2}{2} - b\varepsilon\cos\left(\frac{2\pi u}{a_{Cu}}\right) - p\varepsilon\right], \qquad (6)$$

where Q is the elastic coefficient of the lattice, p is the difference of the pressing force components:

$$\text{p} = \text{p}_1 - \text{p}_2 . \qquad (7)$$

The second term under the integral corresponds to the fact that the system tends to the positive deformation if $u = 0$ (i.e. the ion O is situated on the side of the unit sell parallel to the axis 1) and the negative one if $u = a_{Cu}/2$ (the ion O is situated on the side parallel to the axis 2) (Fig.1). Usually, this term is chosen in a bilinear form (as e.g. in [7]) within the every unit sell. But if we do in this way it will be the jump of force acting on the O ion (the energy derivative in the displacement u) in the vicinity of the points $u = ma_{Cu}/2$, where m is integer. Really, when the O ion is at the right side of the point u = 0 and close enough to this point (see Fig.1), the force will not depend on u. However, when the ion is at the left side, the absolute value of the force will be the same , but with opposite sign. To avoid this unphysical gap we suggest to write the term in the form above.

Thus, when minimizing (6) on ε we obtain

$$\varepsilon = \frac{\text{p}}{Q} + \frac{b}{Q}\cos\left(\frac{2\pi u}{a_{Cu}}\right) . \qquad (8)$$

Substituting (8) into (6) we then find with the accuracy of the constant:

$$H = H|_{\varepsilon=0} + \frac{1}{L}\int_0^L d\phi \left[\frac{b^2}{2Q}\sin^2\left(\frac{2\pi u}{a_{Cu}}\right) - \frac{b\text{p}}{Q}\cos\left(\frac{2\pi u}{a_{Cu}}\right)\right]. \qquad (9)$$

The first item under the integral renormalizes the constant V:

$$V \to V' = V + \frac{b^2}{2Q} . \qquad (10)$$

Thus, if p = 0 the account of elastic interaction of oxygen ions do not change the model, suggested in [1], but renormalizes the corresponding parameters. By going over to the dimensionless values (following [1])

$$\psi = \phi\frac{k\pi\sqrt{2}}{a_{Cu}}, \quad k^2 = \frac{V'}{K}, \quad G = \frac{2}{k}\frac{a_0 - a_{Cu}}{a_{Cu}} \propto \delta, \quad \text{p}' = \frac{2b\text{p}}{QV'}, \quad s = \frac{4\pi u}{a_{Cu}},$$

we obtain instead of (9):

$$H = \frac{1}{L}\int_0^L d\psi \left[-G\frac{ds}{d\psi} + \frac{1}{2}\left(\frac{ds}{d\psi}\right)^2 + 2\sin^2\left(\frac{s}{2}\right) - \text{p}'\cos\left(\frac{s}{2}\right)\right]. \qquad (11)$$

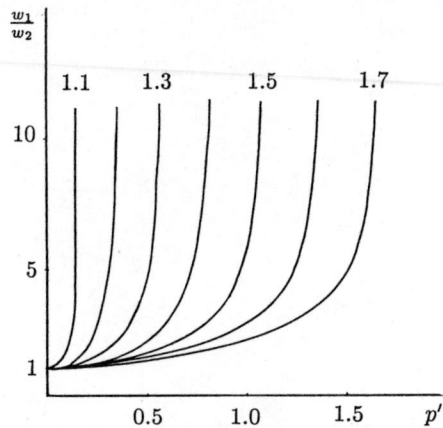

Fig. 2. Dependence of dimension ratio of neighboring twins on pressure. The parameter of the curves is $G\pi/4$

The variational equation for the functional (11) is a stationary double Sin-Gordon [8]:

$$s'' = \sin(s) + (p'/2)\sin(s/2) \ . \tag{12}$$

The latter is easily integrated one time:

$$s' = \pm[2(C - \cos(s)p'\cos(s/2))]^{1/2} \ . \tag{13}$$

The minima of (11) is achieved on the phase trajectory with $s' > 0$, as it is demonstrated in [1]. Such trajectories exist when

$$C > C_m = 1 + |p'| \ . \tag{14}$$

Besides, the trajectory with $C = C_m$ is the separatrice. Minimizing the energy (11) in the period of structure

$$l = \int_0^{4\pi} \frac{ds}{s'(s)} \tag{15}$$

analogous to [1] we obtain that minimum is achieved due the following relation between G, p' and C:

$$G = \frac{1}{2\pi} \int_0^{2\pi} s'(s)ds \ . \tag{16}$$

This integral is simple when $C = C_m$. It determines the critical value of G:

$$G_{\text{cr}} = \frac{4}{\pi}\left[\sqrt{1+x} + x\ln(1+\sqrt{1+x}) - x/2\ln(x)\right] , \tag{17}$$

where $x = |p'|/4$. The twin structure is energetically advantageous when $G > G_{\text{cr}}(p')$ or $\delta > \delta(p')$, $\delta_{\text{cr}} = kG_{\text{cr}}(p')$. This expression is reduced to the extracting from [1] when $p' = 0$.

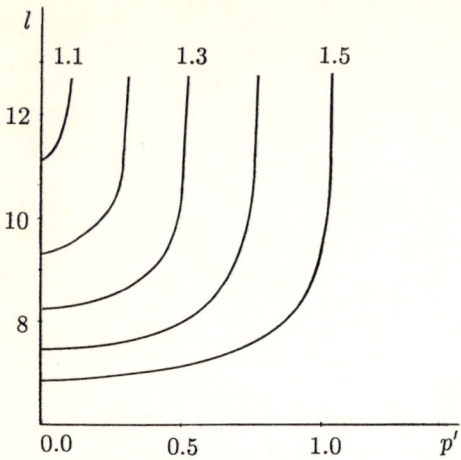

Fig. 3. The dependence of twins period on pressure. The parameter of the curves is $G\pi/4$

As we can see from Fig. 2, the influence of pressure reduces to the compression of only one of the two neighboring domains and expansion of another one. The ratio of dimensions of these twins (the dimension of larger domain) tends to infinity when $p' = p'_{cr}(G)$. This means that twin structure disappears when p become larger than p'. This effect is described also by ferroelastic model of twinning (See [5]). The function $p'_{cr}(G)$ is inverse to that of $G_{cr}(p')$ (17).

The sum of the neighboring domain widths (a period of twin structure) increases with pressure and also tends to infinity when $p' = p'_{cr}(G)$ (Fig. 3). Intertwine boundaries in this case change the thickness insufficiently. The last one may be estimated as the thickness of "walls" of a soliton, being the solution to Eq. (12) when $p' = p'_{cr}(G)$.

The examples of the coordinate dependencies of oxygen atoms displacement u (from correspondent sites in Cu-lattice) and the deformation of Cu-lattice ε are represented in Fig. 4 for p > 0. When p < 0 the twin with smaller deformation is wider.

Finally we can make some conclusions. As follows from the formula (10), the role of the deformations in the twinning process is analogous to the role of the Cu-O interaction. The both interactions "want" to fix O ion at the center between two Cu ions.

The system that described by present model under an external pressure behave like the system described by classical ferroelastic model [2-4], Thus all the ferroelastic effects (including detwinning) can be explained in the frames of our model. But from our model we can obtain the dependencies of equilibrium structure period on different parameters. We point out to the increase of the twin structure period, when the pressure increases. This result can be easy checked experimentally.

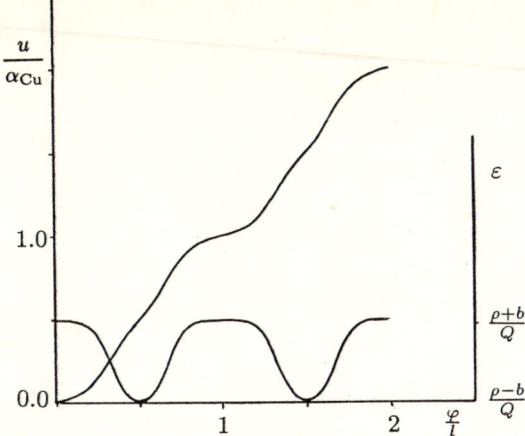

Fig. 4. The example of dependencies of the O-ions displacement u and the lattice deformation ε (periodic curve) on the coordinate ϕ for $p > 0$

The author is grateful to the members of seminar conducted by the academician A.S. Davydov for useful discussions and thankful to the Yu.B. Gaididei and V.M. Loktev for the support.

References

1. Gaididei Yu.B., Loktev V.M., Nikolayev V.S.: Solid State Comm. **75** 503 (1990)
2. Barsch G.R., Krumhansl J.A.: Phys. Rev. Lett. **53** 1069 (1984)
3. Jacobs A.E.: Phys. Rev. **B 31** 5984 (1985)
4. Barsch G.R., Horowitz B., Krumhansl J.A.: Phys. Rev. Lett. **59** 1251 (1987)
5. Schmid H. et al.: Physica **C 157** 555 (1989)
6. Dorosinsky L.A. et al.: JETP (Pis'ma) **49** 501 (1989)
7. Gaididei Yu.B., Loktev V.M.: Sov. Superconductivity **2** 75 (1989)
8. Bullough R.K., Caudrey P.J., Gibbs H.M. in: *Solitons*, Bullough R.K. and Caudrey P.J. (Eds.), (Springer, Berlin – Heidelberg – New-York 1980)

Part II

Correlation Effects in Organic Crystals, Molecules and Polymers

Coexistence of Mott and Peierls Instabilities in Quasi–One–Dimensional Organic Conductors

I.I. Ukrainskii and O.V. Shramko

Institute for Theoretical Physics, Metrologicheskaya 14, 252 130 Kiev, USSR

1. Introduction

The quasi–one–dimensional conductors have so far being studied for a long period (see Review [1]). Also these systems are now of interest for both theoreticians and experimentators (see Review [2]). This interest, on the one hand, is due to advances in synthesis of polyacetylene (PA) polydiacetylene (PDA), organic crystalline conductors based on6 molecular donors and acceptor of electron [2]. On the other hand, one–dimensional (1-d) conductors are nontrivial systems. Thus, 1-d metal is unstable to the transition in semi–conducting state. As a result, the simple 1-d metal with half-filled conduction band becomes the Mott semi–conductor or Peierls semi–conductor [1, 2]. The Peierls transition leads to dimerization – or bond length alternation – of the uniform 1-d lattice and semi–conducting energy gap is proportional to the dimerization amplitude. The Mott transition is a result of electron correlation and energy gap in the Mott semi–conductor vanishes with decreasing electron–electron interaction strength (see Refs. [1, 2] and references there–in). The semi–conductor of the Mott and Peierls type possesses some properties of interest. The Mott semi–conductors are characterized by antiferromagnetic structures [2, 3]. In the Peierls semi–conductors the kink–type excitations are possible [4, 5].

The influence of the Mott and Peierls instabilities on the properties of real quasi–one–dimensional systems is so far being studied.

The main problem in theoretical studies consists in complications of a many–electron approach good enough in large systems. In earlier papers [1] contradiction of the Mott and Peierls transitions was stated. Then it was shown that this contradiction is a result of one–electron – or the SCF–theory. The correct conclusion that the Mott and Peierls transitions coexist one with another was first made by one of these authors in Ref. [6]. This result was obtained due to a correct treatment of pair electron correlations using variational function in geminals product from – varying localized geminals (VLG) approach [7, 8]. In Ref. [6] it is shown also that electron–electron interaction can enhance the Peierls dimerization. This somewhat surprising result initiates the activity of a number of investigators – theoreticians [9–14]. But, the authors of all of the papers [9–14] have obtained the same result – including small electron–electron interaction leads to the increase in dimerization. This conclusion was made from a very precise treatment of short chains with allowance for electron correlations. Then, this result has been received on the basis of perturbation theory for infinite chains using computers [12] and the Feynman diagram technique [13]. The authors of Ref. [14] performed numeric calculations of short chains with open and within the same geminals approach as in Ref. [6] and also obtained the same results slightly deformed by boundary conditions.

Thus, we can state now that the theory predicts coexistence if Mott and Peierls instabilities in real systems. So, the experimental data on 1-d system should not correspond to the simple picture if the Peierls semiconductor or the Mott semiconductor [1]. We must explore the more complete theoretical model including the both phenomena. On this way only we can give correct descriptions of real 1-d materials. For example we can give the correct answer to the question what mechanism of the forbidden gap formation is more essential – the electron correlations or dimerization.

In the present paper we study the simultaneous effects of the Mott and Peierls instabilities on electronic spectra and lattice distortion in real 1-d system such as organic donor acceptor molecular crystals and conjugated polymers of PA type. Our studies is based on the varying localized geminals (VLG) approach, described elsewhere [6–8].

2. The method of calculations and qualitative evaluations

Studying the electronic properties if organic 1-d materials we use the model of uniform chain with the adiabatic Hamiltonian

$$H = \sum_{\sigma,m=1}^{N} \beta_m (c^+_{m,\sigma} c_{m+1,\sigma} + c^+_{m+1,\sigma} c_{m,\sigma})$$
$$+ \gamma \sum_m c^+_{m\uparrow} c_{m\uparrow} c^+_{m\downarrow} c_{m\downarrow} + \gamma_1 \sum_m n_m n_{m+1} \qquad (1)$$
$$+ \frac{K_\sigma}{2} \sum_m (x_m - x_{m+1})^2 \equiv T + V_{e-e} V_{e-\text{ph}} ,$$

where $c^+_{m\sigma}$ is the creation operator of electron with σ - spin at m-th site, $n_{m\sigma} = c^+_{m\sigma} c_{m\sigma}$, $N \to \infty$ is a number of sites, x_m is the m-th site displacement. The resonance integrals

$$\beta_m = -[\beta + (x_{m+1} - x_m)\beta'] = -\beta(1 + \Delta_m) , \qquad (2)$$

$\beta', \beta > 0$, K_σ is the lattice elasticity constant, γ and $\gamma_1 > 0$ are the electron repulsion parameters.

We restrict our treatment by the most interesting case of half–filled conduction hand, so, the electron number $N_e = N$. The Peierls deformation in this case reduces to the chain dimerization

$$x_{m+1} - x_m = (-1)^m x_0 , \quad \beta_m = -\beta[1 + (-1)^m \Delta] . \qquad (3)$$

The real experimental values of displacements are small as compared to the lattice constant a. In polyacetylene $x_0 = 0.07\text{Å}$ and $a = 1.395\text{Å}$, [15, 3], for K-TCNQ complexes $x_0 = 0.18\text{Å}, a = 3.6\text{Å}$ [16]. For small values if x_0 the linear dependence is valid

$$\Delta = (\beta'/\beta)x_0 . \qquad (4)$$

The increase in displacements $x_0 \to a$ destroys the relation (4) and harmonic adiabatic approach used in (1). So, the method used here is valid only for small values of $\Delta \ll 1$. In this region Hamiltonian (1) is the Frohlich–type Hamiltonian with linear, relative to displacements x_m, electron–phonon interaction.

So, when $\Delta \ll 1$ the adiabatic approach is good enough and the problem of 1–d instabilities is reduced to studying the ground state energy dependence on the value of Δ (4). Thus, we need the Δ-value, optimizing the expression

$$\varepsilon_t(\Delta) = \varepsilon_{\text{el}}(\Delta) + \frac{1}{2}\Delta^2/\kappa , \qquad (5)$$

where ε_{el} is electron contribution into the ground state energy per an electron pair, the dimensionless constant of electron–phonon interaction.

$$\kappa = (\beta')^2/(2K_\sigma \beta) . \qquad (6)$$

In order to calculate electronic contribution of energy we shall use the VLG approach [6–8]. The ground state wave function has the form

$$\Psi_0 = \prod_{m=1}^{M} G_m^+ |0\rangle \equiv \prod_{m=1}^{M} (u f_{m\uparrow}^+ f_\downarrow^+ + v \tilde{f}_{m\uparrow}^+ \tilde{f}_{m\downarrow}^+)|0\rangle , \qquad (7)$$

where

$$f_{m\sigma} = \sqrt{\frac{2}{N}} \sum_{|k|<K_F} A_{k\sigma} e^{-ikR_m}, \quad \tilde{f}_{m\sigma} = \sqrt{\frac{2}{N}} \sum_{|k|<K_F} \tilde{A}_{k\sigma} e^{-ikR_m} , \qquad (8)$$

$$\begin{aligned} A_{k\sigma} &= a_{k\sigma} \cos\theta_k + a_{\bar{k}\sigma} i \sin\theta_k, \\ \tilde{A}_{k\sigma} &= a_{\bar{k}\sigma} \cos\theta_k + a_{k\sigma} i \sin\theta_k , \end{aligned} \qquad (9)$$

$$a_{k\sigma} = \frac{1}{\sqrt{N}} \sum_{n=1}^{N} c_{m\sigma} e^{-ikna} , \qquad (10)$$

$$u = \cos\varphi, \quad v = \sin\varphi , \qquad (11)$$

$$2\theta_k = \arctan(\lambda \operatorname{tg} ka)$$
$$k = 2\pi l/Na, \quad (l = 0, \pm 1, \pm 2, \ldots), \tag{12}$$

φ and λ are the variational parameters, the Fermi operators $f_{m\sigma}$ and $\tilde{f}_{m\sigma}$ correspond to the orbitals f_m and $\tilde{f}_m$ which are partially localized near points

$$R_m = (2m + \delta)a . \tag{13}$$

The ground state energy in β–units per electron pair has the form

$$\varepsilon_{\text{el}} = 2t\cos 2\varphi - k_0 \sin 2\varphi - v_1 \left(\frac{2}{N} \sum_l |P_l|^2\right) \cos^2 2\varphi , \tag{14}$$

where the kinetic energy average

$$t = \sum_m [1 + (-1)^m \Delta] f_m(n) f_m(n+1)$$
$$= \langle 0 | f_{m\sigma} T f_{m\sigma}^+ | 0 \rangle , \tag{15}$$

$$T = \sum_m (c_{m,\sigma}^+ c_{m+1,\sigma} + c_{m+1,\sigma}^+ c_{m,\sigma}) , \tag{16}$$

the exchange integral

$$K = \langle 0 | f_{m\uparrow} f_{m\downarrow} V_{e-e} f_{m\uparrow}^+ f_{m\downarrow}^+ | 0 \rangle =$$
$$= U \sum_n |f_0(n)|^4 - U_1 \sum_n |f_0(n)|^2 \cdot |f_0(n+1)|^2 , \tag{17}$$
$$U = \gamma/\beta, \quad U_1 = \gamma_1/\beta ,$$

average of non diagonal density or bond order

$$P_l = \langle \Psi_0 | c_{l,\sigma}^+ c_{l+1,\sigma} + c_{l+1,\sigma}^+ c_{l,\sigma} | \Psi_0 \rangle =$$
$$= \sum_m f_m^*(l) f_m(l+1) = \left[\frac{1}{\pi} + (-1)^l \frac{\lambda}{4\pi} \ln \lambda\right] \cos 2\varphi . \tag{18}$$

Non we consider the Hubbard approach $\gamma_1 = 0$ in (1). Then, variation of the energy (14) with respect to φ gives

$$\varepsilon_{\text{el}} = -\varepsilon_g + U/2 , \tag{19}$$

where

$$\varepsilon_g = \sqrt{4t^2 + K^2} . \tag{20}$$

The values of t, K, p depend on the value of λ in Ref. [12, 6], so

$$t(\lambda) = -\frac{4}{\pi}\left[E(1-x^2) + (4-\lambda)\frac{\partial E(1-\lambda^2)}{\partial \lambda}\right], \tag{21}$$

where the $E(x)$ is the elliptic integral.

The explicit form of λ–dependence on K can be obtained in the limit of small

$$K(\lambda) = \frac{1}{3}U - \mathrm{const} \cdot \lambda \cdot \ln \lambda \;. \tag{22}$$

We can see from (22) that when λ and, as a result, U are small the energy dependence (20) on λ is nonanalytic. Thus, we can suppose strong U–dependence of Δ_0 which minimizes the total energy. In the next section we study numerically the U–dependence of λ. Here we evaluate asymptotic behaviour in the limiting cases $U \to 0$ and $U \to \infty$.

When $U \to 0$ the non–interacting electron model is valid and the energy is defined by the value of (21) and its optimization with respect to λ gives $\lambda = \Delta$. The energy minimum corresponds to

$$\Delta_0 = 4\exp\{-\pi/8\kappa\}_{(U\to 0)} \tag{23}$$

due to the fact that [6]

$$\varepsilon_t = 2\left(-\frac{4}{\pi} - \frac{2\Delta^2}{\pi}\ln\frac{4}{\Delta}\right) + \frac{1}{2\kappa}\Delta^2 \;. \tag{24}$$

When $U \gtrsim 4$ we can use the simpler approach instead of (12)

$$\theta_k = \widetilde{\lambda} k \;. \tag{25}$$

Using (25) we obtain [7]

$$\widetilde{t}(\lambda) = -\frac{4}{\pi}\frac{\cos\pi\widetilde{\lambda}}{(1 - 4\widetilde{\lambda}^2)} \;. \tag{21a}$$

$$\widetilde{K}(\lambda) = \frac{U}{3}\left(1 + \frac{1}{2}\sin\pi\lambda\right) \;. \tag{22a}$$

Substituting (21a, 22a) in (19) and (5) and optimizing Δ we obtain

$$\Delta_0 = \frac{8\kappa}{U}\left(1 - \frac{4}{U^2}\right) \;. \tag{26}$$

We note that the (21a, 22a, 19) describe good enough the U–dependence of the total energy for any values of $U > 0$ [7]. But, the correct description of the Peierls instability hear the point $U = 0$ needs more precise relations due to the fact that the Peierls instability results from a logarithmic term. The latter is just lost when passing from (12) to (25), as we noted in Ref. [6]. Now we consider the effect of electron–electron interactions at neighboring sites, resulting from terms γ_1 in (1). Studying (14, 18) we can conclude that γ_1–term increases the amplitude of dimerization. In the limiting case of small interactions $U_1 < U \to 0$ we obtain

$$\Delta_0(U_1) = \Delta_0(U_1 = 0) \cdot \exp[\frac{\pi U_1}{24\kappa^2}] = 4\exp[-\frac{\pi}{8\kappa}] \cdot \exp[\frac{\pi U_1}{24\kappa^2}] \;. \tag{27}$$

So, we can see an exponential increase of Δ_0 with $U_1 > 0$.

3. Numeric Calculations and Chain–Length–Dependence

In order to define the optimal value of Δ_0 we must seek for the minimum of the energy (5) taking into account (19) in the space of Δ_0 and λ variables namely

$$E(\lambda, \Delta, U) = -[4t^2(\lambda, \Delta) + K^2(\lambda, U)]^{1/2} + \frac{1}{2\kappa}\Delta^2, \qquad (28)$$

where t and K are defined respectively by (15, 17). This problem is not too complicated as far as all the necessary values can be written in quadrature (14–18). But, when $U \ll 1$ some technical difficulties arise with the increase in chain length due to the logarithmic Δ-dependence of the electronic energies in (19, 24). As a result we cannot use the standard method of quantum–chemical optimization of the bond lengths. This method is based on the linear relations between bond length and bond order resulting from energy expansion

$$\varepsilon(\Delta) = \varepsilon_0 + \varepsilon' \cdot \Delta + \frac{1}{2}\varepsilon'' \cdot \Delta^2,$$

where

$$\varepsilon' \approx \frac{1}{N}\beta' \sum_l P_l(x_l - x_{l-1}), \quad \varepsilon'' \approx K_\sigma$$

and, as a result,

$$\Delta_0 = \frac{1}{N}\beta' P_l / K_\sigma. \qquad (29)$$

But we can see from (24) that the first derivative of $\varepsilon_t(\Delta)$ vanishes at the point $\Delta = 0$ and the second one diverges at this point. So, the calculations based on the assumption that linear dependence (29) and energy expansion in series of Δ (29) do exist really do not deal with real Peierls instability. First this difficulty was mentioned in Ref. [1].

Some calculations on short PA–chains are based on the formula (29). The logarithmic contribution in energy of short chains $N < 50$ is, certainly, absent. But, on the other hand, the results of such calculations do not reproduce the correct values of the Peierls dimerization Δ in infinite 1-d system. These results are characteristics of the model chain with small lengths themselves only, depending also on the type of boundary conditions cyclic chains, open chains, etc. This dependence vanishes when N is large enough and we deal with the Δ-dependencies if the form (23, 24).

Table 1. Mean kinetic energy t in chains of different lengths, $\Delta = 0$, $U = 0$, $\lambda = 0$

N	t_N	$\Delta = t_N - t_\infty$	$\ln N$	$1/N$
6	1.333333	0.060043	1.79	0.16667
30	1.275570	0.002280	3.40	0.03333
70	1.273667	0.000377	4.25	0.01428
202	1.273291	0.0	5.31	0.00495
∞	1.273290	0.0		

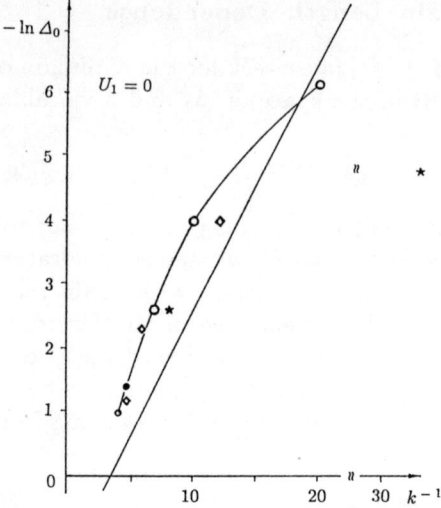

Fig. 1. The dependence of the Peierls deformation Δ_0 amplitude on the e-ph interaction parameter (6) $k = \kappa$ for non-interacting electrons: • – Hirsch, 1983, $N = 24$ [11]; ◇ – Baeryswyl & Maki, 1985, $N = \infty$ [12]; ⋆ – Kuprievich, $N = 30$, [14]; ○– present paper, (——) – expression (23), $N = \infty$; ($\cdot - \cdot - \cdot$) – numerical minimization of (28) obtained by Ukrainskii in 1979 [6]

The Table 1 illustrates the above statements with the N – dependence of the conduction bend width or the mean values of kinetic energies. As we can see even for comparatively long chains with $N = 70$ the difference between t_N and t_∞ is few units of 10^{-4} order. The same order is also the Peierls contribution into the ground state energy $\Delta^2 \ln \Delta$ when $\Delta \lesssim 0.01$.

When $U > 1$ the problem is simpler due to the transformation of Δ –dependence of the electronic energy [19].

Now we consider the numerical results. In Figs. 1–3 we give dependencies of dimerization Δ_0 in (28) on electron–electron interaction strength and electron–photon interaction parameter κ.

For minimization of the total energy we use the direct search of extreme procedure [18] by means of (6–19) which are valid for cyclic chains with any $N = 4n + 2$. So we can compare our results with calculations of short [9, 10, 11, 14] and infinite [12, 13] chains.

We can see from Fig. 1 that in results of calculations of short cyclic chains [9, 10, 11, 12] the dimerization amplitude is underestimated as compared to exact value for infinite chains (23) for the region $\Delta > 0.05$ and overestimated when $\Delta < 0.05$. The largest overestimation of Δ especially for most actual region $\kappa < 0.1$ is connected with using open chain model [14]. This fact shows once again that the boundary conditions are of importance, in the case of short chain models.

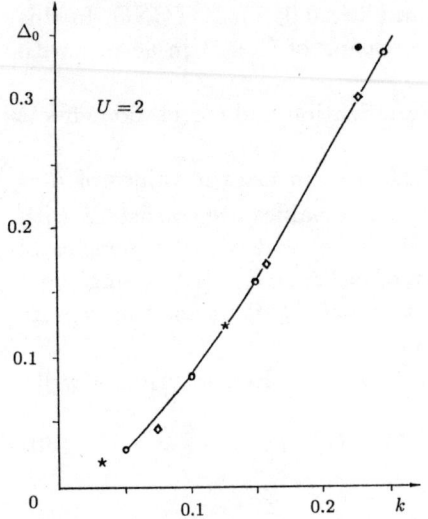

Fig. 2. The $k = \kappa$-dependence of dimerization amplitude Δ_0: • – Hirsch, 1983, [11]; ◊ – Baeryswyl & Maki, 1985, [12]; ⋆ – Kuprievich, [14]; present calculation with expression (28)

The κ-dependence of Δ_0 for $U = 2.4$ is given in Figs. 2, 3 where the difference between various approaches decreases with increasing of U.

The Fig. 4 shows the U–dependence of Δ_0. In the region of large $U > 4$ a good approach can be obtained analytically be (26). The values of $\kappa > 0.3$ are non–realistic and cannot be treated by the linear Frohlich models. In

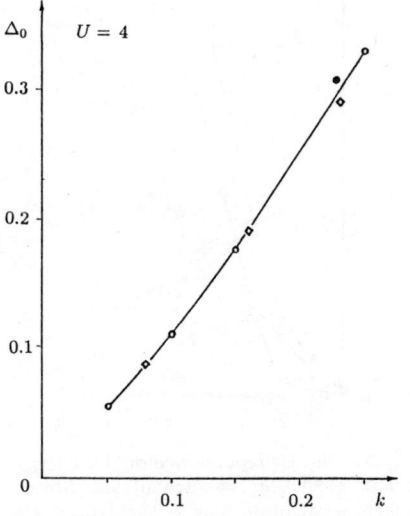

Fig. 3. The $k = \kappa$-dependence of dimerization amplitude Δ_0. See Fig. 2 for notations

the real system we have $\kappa = 0.07$ for PA and $\kappa = 0.05$ for K-TCNQ. In this region of κ-values we can rise Δ with increasing of $U \ll 1$ in accord with results of Ref. [6].

Now we consider the contribution of dimerization and correlation effects in optical spectra of organic materials.

In conjugated polymers like PA or PDA we can use the values of $\beta = 2.4\,\text{eV}$, $\beta' = 4\,\text{eV/Å}$, $K_\sigma = 47\,\text{eV/Å}^2$ [19]. These values are consistent with the data on small conjugated molecules [1, 2, 19] and with frequencies of vibrations active in IR and Raman spectra of of PA $(CH)_x$ [19]. Using these values of parameters and (6) we obtain that $\kappa = 0.07$. It means that we are in the region of strong U-dependence of Δ_0.

Now we calculate the dielectric gap ΔE. Due to Ref. [8] we can write down

$$\Delta E = 2[\varepsilon_g - t_0(1 + U^2) + T_k U^2], \qquad (30)$$

where T_k is conduction electron energy

$$T_k = \sum_m e^{ikm} \langle f_n | \widetilde{T} | f_{n+m} \rangle .$$

The gap value ΔE_p (30) consists of correlation contribution ΔE_{corr} and dimerization contribution ΔE_{dim}. When U is small we can assume

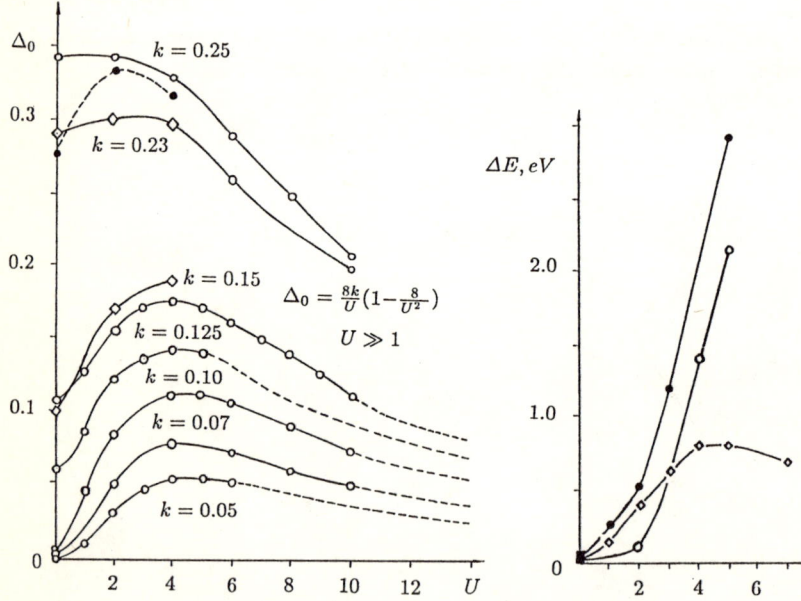

Fig. 4. The U-dependence of dimerization amplitude. See Fig. 2 for notation

Fig. 5. The U-dependence of the energy gap in electronic spectra of the Mott–Peierls semiconductor: $\diamond$ – ΔE_{dim}; $\circ$ – ΔE_{corr}; $\bullet$ – ΔE

$$\Delta E_{\text{corr}} = 2\varepsilon_g - 2t_0, \quad \Delta E_{\text{dim}} = 4\beta\Delta_0, \tag{31}$$

where ε_g is determined by (20), t_0 – by (21a).

Fig.5 shows the U–dependence of values in (31). It follows from Fig.5 that then $U \lesssim 1.5$ the dimerization contributions ΔE_{dim} exceed the correlation contribution ΔE_{corr}. This fact is due to the strong U–dependence of Δ_0.

4. Comparison with the Experiment

Now we reevaluate the parameters of real organic conductors using U–depen- dence of ΔE from Fig.5.

Studying experimental data on trans–PA we can conclude that $\Delta E = 1.9\,\text{eV}$ [2, 19]. Using the above evaluation $\kappa = 0.07$ we obtain $U = 2.5$ so $\gamma = 6.2\,\text{eV}$. It is of interest to note that in this region $\Delta E_{\text{corr}} \lesssim \Delta E_{\text{dim}}$. Nearly the same situation occurs in PDA, where $\Delta E = 2.5\,\text{eV}$.

Now using Fig.5 we can easily understand why there are some differences in evaluations of correlation and dimerization contributions in ΔE–value. Namely in the region of intermediate values of $1 < U < 4$ the correlation part ΔE_{corr} sharply increases, exceeding ΔE_{dim}. So, the values of $U, \kappa < 0.1$ are strongly dependent on small perturbations such as boundary conditions or chain lengths.

In organic materials like K-TCNQ we have instead $\Delta E = 0.9\,\text{eV}$, $\kappa = 0.05$, $t = 0.15\,\text{eV}$ so $U \gtrsim 6$. Thus, the correlation contribution in ΔE is dominant and we can use (26) for the evaluation of Δ_0.

Also, we can conclude that the agreement of the calculated values of Δ_0 or x_0 in (4) can be obtained in different approaches. But it needs different values of parameters γ, K_σ, β, β' which depend on the model used in calculations of short or infinite chains and also cyclic boundary conditions. Giving preference to either calculation model we must bare in mind a number of experimental data, not only the values of Δ_0. Our VLG approach [6–8] allows us to explain the experiments on Δ_0 and ΔE in $(CH)_x$ and K-TCNQ materials vibration frequencies in $(CH)_x$ using the values of parameter of Hamiltonian (1) which in the case of PA are in common use for conjugated molecular systems.

References

1. Ovchinnikov A.A. Ukrainskii I.I., Kventsel G.F.: Uspechi Fiz. Nauk **108** 81 (1972) [Soviet Phys. Uspekhi **15** (1973)]
2. Ovchinnikov A.A. Ukrainskii I.I.: Soviet Sci. Rev. Section B Chem. Rev., Volpin M.E. (Ed.) **9** 123 (1987)

3. Ovchinnikov A.A.: Theor. Chim. Acta (Berlin) **47** 297 (1978)
4. Su W.P., Schrieffer J.B., Heeger A.J.: Phys. Rev. **22** 2099 (1979)
5. Brazovskii S.A.: J. Exp. Theor. Phys. **78** 677 (1980)
6. Ukrainskii I.I.: J. Exp. Theor. Phys **76** 760 (1979) [Sov. Phys. JETP **49** 381 (1979)]
7. Ukrainskii I.I.: Teor. Matem. Fiz. **32** 392 (1977), TMP **32** 816 (1978)
8. Ukrainskii I.I.: Phys. Stat. Solidi **106** 55 (1981)
9. Mazumbar S., Dixit S.N.: Phys. Rev. Lett. **51** 292 (1983)
10. Dixit S.N., Mazumbar S.: Phys. Rev. **B 29** 1824 (1984)
11. Hirch: Phys. Rev. Lett. **51** 296 (1983)
12. Baeryswyl D., Maki K.: Phys. Rev. **B 31** 6633 (1985)
13. Krivnov V.Ya., Ovchinnikov A.A.: J. Exp. Theor. Phys. Lett. **39** 134 (1984)
14. Kuprievich V.A.: Teor. Exper. Khimiya **22** 263 (1986)
15. Fincher C.R., Chen C.-E., Heeger A.J., Macdiarmid A.G., Hastings J.B.: Phys. Rev. Lett. **48** 100 (1982)
16. Hoekstra A., Spoelder T.: Vos A. Acta Crystallograph. **B 28** 14 (1972)
17. Ukrainskii I.I.: ITP-81-173E-Preprint, Kiev, 1984; *Physics of Many Particle Systems* (Naukova Dumka, Kiev 1988) **13** 89
18. Wild D.J.: *Search of Extrema* (Oxford 1989)
19. Ochinnikov A.A., Belinskii A.E., Misurkin I.A., Ukrainskii I.I.: ITP-81-18E-Preprint, Kiev; Internat. J. Quantum Chem. **22** 761 (1982)

Nonlinear Optical Susceptibility for Third Harmonic Generation in Combined Peierls Dielectrics

Yu.I. Dakhnovskii and K.A. Pronin

Institute of Chemical Physics, Kosygin St. 4, 117977 Moscow, USSR

With the use of the nonequilibrium Keldysh diagram technique we calculate the nonlinear optical susceptibility of combined Peierls dielectrics for the process of third harmonic generation. Slow relaxation of carriers is taken into account. The frequency dependence of the susceptibility has a resonance structure. The square root divergencies at frequencies corresponding to one-third and the total band gap are smeared out by damping.

1. Introduction

The nonlinear optical properties of conjugated polymers are of considerable interest [1–3]. These substances form a new class of perspective materials exhibiting very high nonlinearities. For third harmonic generation the experimentally measured values of the susceptibility $\chi^{(3)}$ are of the order $\sim 10^{-11}$ e.s.u. for polydiacetylene [4], $\sim 10^{-10}$ e.s.u. for cis-$(CH)_x$ and polythiophene [5, 6] and $\sim 10^{-8} - 10^{-9}$ e.s.u. for oriented trans–polyacetylene [7, 8]. Besides, the values of the dielectric constants for these polymers are sufficiently low. All these features make conjugated systems good candidates for practical applications in fast optical switches, waveguides and optical harmonic generators.

There is a large number of publications devoted to the study of electrical, magnetic and optical properties of organic semiconductors [9, 10]. The theory of nonlinear optical effects in polymers is not so rich with rigorous results [1–3, 11, 12]. Especially we note here the theoretical paper by Wu [11], where the nonlinear susceptibility of a Peierls dielectric with a dimerised degenerate ground state (that corresponds to trans-$(CH)_x$ without solitons) has been rigorously calculated without taking into account the coulomb interactions and relaxation processes. The frequency dependence of $\chi^{(3)}$ has been found to be in good agreement with experimental data for trans-$(CH)_x$.

In this paper we calculate the nonlinear optical susceptibility of a combined Peierls dielectric for the process of third harmonic generation. We take into account the slow relaxation of carriers, originating from inelastic electron scattering on thermal phonons and impurities.

2. The Model

We consider one–dimensional combined Peierls dielectrics [10]. As examples we mention polydiacetylene, cis-$(CH)_x$, polyphenylene and other conjugated polymers except for trans-$(CH)_x$. These substances are semiconductors (i.e. they have a gap in the spectrum) due to their chemical structure even if we do not take into account the interaction between the electrons and the lattice. At the same time the Peierls effect is of considerable importance and must be accounted for. Here we will use the one–particle approximation with respect to electrons.

The continuum Hamiltonian of a combined dielectric has the form

$$H = \int dx\, \Psi^+(x) \left\{ -i\hbar v_f \sigma_3 \frac{d}{dx} + \Delta^* \sigma_+ + \Delta \sigma_- \right\} + \int dx\, g^{-2} \Delta_p^2(x)\,. \tag{1}$$

Here $\Psi^+(x) = [\Psi_1^+(x), \Psi_2^+(x)]$ is a wave function of a spinor type corresponding to the two sorts of electrons (left– and right– going) that evoked after the linearization of the electron spectrum in the derivation of (1) [9, 10]; $\sigma_\pm = \frac{1}{2}(\sigma_1 \pm \sigma_2)$, σ_i are the Pauli matrices acting in the space of indices labelling the types of electrons; δ is the complex gap

$$\Delta = \Delta_s + \Delta_p e^{i\zeta}\,. \tag{2}$$

Its component Δ_s corresponds to the interaction of the carrier with the initial rigid structure of the molecule (the effect of coulomb correlations in the mean field approximation can be incorporated in Δ_s) while Δ_p having a phase shift ζ originates from the Peierls dimerization.

Now we introduce the external electric field in the following calibration

$$\varphi = 0\,, \qquad A = -E_0 c \omega^{-1} \sin \omega t\,. \tag{3}$$

The "zero" Hamiltonian is represented by the electron part of (1), and the interaction Hamiltonian is defined by

$$\Delta H = -c^{-1} \int dx\, j(x,t) A(t)\,, \tag{4}$$

where j is the current operator

$$j(x,t) = e v_f \Psi^+(x) \sigma_3 \Psi(x)\,. \tag{5}$$

3. The Nonlinear Susceptibility

Our goal is to find the nonlinear susceptibility defined through the expansion of the polarization in powers of the applied field

$$\langle P \rangle = \chi^{(1)} E + \chi^{(2)} EE + \chi^{(3)} EEE + \ldots. \tag{6}$$

Systems possessing inversion symmetry have vanishing $\chi^{(2)}$. So we calculate the nonlinear response in the lowest third order process. As we consider a time–dependent problem with a harmonic external field having the frequency comparable to the band gap, we use the Keldysh nonequilibrium diagram technique [13, 11]. Then the time integration and the T – operator permutations are performed on the Keldysh contour t_c consisting of two branches: 1– going from $-\infty$ to $+\infty$ and 2 - the reverse from $+\infty$ to $-\infty$.

We calculate the expansion of the average current

$$\langle j(x,t) \rangle = \frac{d}{dt} \langle P(x,t) \rangle$$

in powers of the applied field. According to Keldysh rules for the correlation functions [13], in order to preserve the right sequence of operators in $j(x,t)$ we assign to $\psi^+(x,t_2')$, $\psi(x,t_1')$ equal times $t = t_1' = t_2'$ that are placed, however, on different branches of the contour. It means that in our consideration we should calculate the four current correlation function with external Keldysh indices 1, 2 fixed.

$$\chi^{(3)}(\Omega; \omega_1, \omega_2, \omega_3) = \frac{i}{3! \hbar^3 \Omega \omega_1 \omega_2 \omega_3} \int_{t_c} dt_1 dt_2 dt_3 \exp(i \Sigma \omega_i t_i)$$
$$\times \int dt \exp(i\Omega t) \int dx_1 dx_2 dx_3 dx \exp(-iQx) \tag{7}$$
$$\times \langle T_c j(x, t = t_1' = t_2') j(x_1, t_1) j(x_2, t_2) j(x_3, t_3) \rangle .$$

After performing the pairing procedure we obtain the formula that corresponds to the sum of six identical diagrams. Each diagram represents a loop constructed of electron Green functions (GF) with four vertices corresponding to the interaction with the external field

$$\chi^{(3)}(\Omega; \omega_1, \omega_2, \omega_3) = \frac{i\hbar(ev_f)^4}{2\pi \Omega \omega_1 \omega_2 \omega_3} \delta(\Omega + \omega_1 + \omega_2 + \omega_3)$$
$$\times \int dk \int dv\, \sigma_3 G(v) \sigma_3 \tau_3 G(\omega_1 + v)$$
$$\times \sigma_3 \tau_3 G(\omega_1 + \omega_2 + v) \sigma_3 \tau_3 G(\omega_1 + \omega_2 + \omega_3 + v) . \tag{8}$$

Here the matrix products with respect to the upper Keldysh indices not written out explicitly (Pauli matrices τ_3) and the indices of the two types of particles (σ_3 matrices) are implied.

In the zero GF we phenomenologically introduce weak damping. Thus we take into account the relaxation processes, originating from electron scattering on thermal phonons and disorder not incorporated in the Hamiltonian.

$$G^{11}(k,v) = \hbar^{-1}\frac{v + kv_f\sigma_3 + \Delta^*\hbar^{-1}\sigma_+ + \Delta\hbar^{-1}\sigma_-}{v^2 - (\omega_k - i\varepsilon)^2} , \qquad (9)$$

$$G^{22}(k,v) = -\hbar^{-1}\frac{v + kv_f\sigma_3 + \Delta^*\hbar^{-1}\sigma_+ + \Delta\hbar^{-1}\sigma_-}{v^2 - (\omega_k + i\varepsilon)^2} , \qquad (10)$$

$$G^{12}(k,v) = \frac{i\varepsilon(v + 2\omega_k)}{\hbar\omega_k^2}\frac{\omega_k + kv_f\sigma_3 + \Delta^*\hbar^{-1}\sigma_+ - \Delta\hbar^{-1}\sigma_-}{(v + \omega_k)^2 + \varepsilon^2} , \qquad (11)$$

$$G^{21}(k,v) = \frac{i\varepsilon(v - 2\omega_k)}{\hbar\omega_k^2}\frac{\omega_k + kv_f\sigma_3 + \Delta^*\hbar^{-1}\sigma_+ + \Delta\hbar^{-1}\sigma_-}{(v - \omega_k)^2 + \varepsilon^2} . \qquad (12)$$

The exact zero GF of the Hamiltonian (1) can be obtained from (9–12) by taking the limit $\varepsilon \to 0$. Then the Lorenzians in (11,12) transform into δ-functions. Imaginary correction terms $\sim \varepsilon$ in the denominators of (9, 10) introduce exponential relaxation with rate constant ε (we take it independent of k,v). Equations (11, 12) are obtained from (9, 10) with the help of the following relations between the "dressed" GF:

$$G^{11}(x_1,x_2) = \theta(t_1 - t_2)G^{21}(x_1,x_2) + \theta(t_2 - t_1)G^{12}(x_1,x_2) ,$$

$$G^{22}(x_1,x_2) = \theta(t_2 - t_1)G^{21}(x_1,x_2) + \theta(t_1 - t_2)G^{12}(x_1,x_2) .$$

The subsequent calculations are performed rigorously for $\hbar\varepsilon \ll |\Delta|$. The results for the Hamiltonian (1) are exact. The microscopic justification for the introduction of relaxation in the form of (9–12) will be given in future publications.

Consider the third harmonic generation process, $\omega_1 = \omega_2 = \omega_3 = \omega$. Performing the summations, taking the integrals of the pole expressions over v and over k we arrive after cumbersome calculations to the final result

$$\chi^{(3)}_{THG} = Bz^{-8}\{3(1 - 8z^2)f(3z) - 8(1 - 4z^2)f(2z) + (5 - 8z^2)f(z)\} . \qquad (13)$$

Here $z = \hbar\omega/2|\Delta|$ is the photon energy divided by the gap width; $B = 3^{-2}2^{-7}e^4(\hbar v_f)^3|\Delta|^{-6}$, and the function f is defined by

$$f(nz) = g(nz - i\varepsilon\hbar|\Delta|^{-1}) - g(-nz) , \qquad (14)$$

$$g(z) = \frac{\ln[\sqrt{1-z} - i\sqrt{1+z}] - \ln\sqrt{2}}{z\sqrt{1-z^2}} . \qquad (15)$$

In the domain $1 - nz \gg \alpha$, $\alpha = \hbar\varepsilon/|\Delta|$ we can neglect weak damping in $f(nz)$ and obtain

$$f(z) = \frac{\arcsin(z)}{z\sqrt{1-z^2}} . \qquad (16)$$

Similarly for $nz - 1 \gg \alpha$ we have

$$f(z) = \frac{1}{z\sqrt{z^2 - 1}} \left\{ \frac{i\pi}{2} - \ln\left[z + \sqrt{z^2 - 1}\right] \right\}. \qquad (17)$$

In the absence of damping the function f(nz) exhibits a square root divergence in the vicinity of $nz = 1$ (16, 17). Keeping the relaxation corrections we obtain the main terms with respect to α

$$f(z) = \frac{\pi}{4} \alpha^{-1/2} \frac{\sqrt{\sqrt{1+y^2} - y} - i\sqrt{\sqrt{1+y^2} + y}}{\sqrt{1+y^2}} - 1, \qquad (18)$$

where $y = \Delta(z-1)/\hbar\varepsilon$, $|y| \lesssim 1$.

Now we analyze the obtained general expressions for $\chi^{(3)}$ (13–15). Far away from the resonances the frequency dependence of the susceptibility calculated by us for the combined dielectric model with weak relaxation coincide in the main terms with the result for the purely Peierls dielectric without damping [11]. Drastic changes arise in the vicinity of the frequencies $z = 1, 1/3$. Here the square root divergencies are now smoothed out by damping, but because of the inequality $\hbar\varepsilon \ll |\Delta|$ the resonance structure is still highly pronounced. We consider in detail the $z = 1/3, 1/2$ resonances as this frequency range is of practical interest for typical polymers.

For $|z - \frac{1}{3}| \ll \alpha$ we have

$$\chi^{(3)}_{\text{THG}} = B3^8 \left\{ \frac{\pi(1-i)}{12\alpha^{1/2}} - a - \left(z - \frac{1}{3}\right) \left[\frac{\pi(1+i)}{8\alpha^{3/2}} + \frac{6\pi(1-i)}{\alpha^{1/2}} - b \right] + \cdots \right\},$$

$$a = \frac{1}{3} + 4 \cdot 5^{1/2} \arcsin\left(\frac{2}{3}\right) - 37 \cdot 2^{-3/2} \arcsin\left(\frac{1}{3}\right), \qquad (19)$$

$$b = 24a + 5\frac{7}{8} + 108 \cdot 5^{-1/2} \arcsin\left(\frac{2}{3}\right) - 1161 \cdot 2^{-9/2} \arcsin\left(\frac{1}{3}\right).$$

For $z \sim 1/3$ this linear function takes finite values inversely proportional to small $\alpha^{1/2}$.

Below the resonance frequency $\alpha/3 \lesssim 1/3 - z \ll 6^{-3}\pi^2\alpha^{-2}$, $\chi^{(3)}$ is given by the formula

$$\chi^{(3)}_{\text{THG}} = Bz^{-8} \left\{ \frac{\pi}{6^{3/2}(1/3 - z)^{1/2}} - a - \frac{i\pi\alpha}{6^{5/2}(1/3 - z)^{3/2}} + \cdots \right\}. \qquad (20)$$

Equation (20) transforms smoothly into the asymptotes that follow from (13, 16) for the case of zero damping. The real part of the latter exhibits a square root divergence given by the first term in figure brackets in (20). Now $\chi^{(3)}$ has a peak structure with the maximum values of both the real and imaginary parts proportional to $\alpha^{-1/2} \gg 1$. We also note that damping produces a nonzero tail for $\text{Im}\chi^{(3)}$ in the resonance area $z < 1/3$, where for

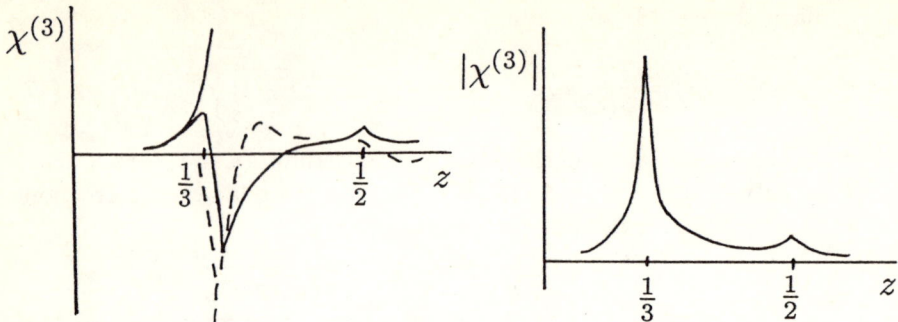

Fig. 1. The nonlinear susceptibility. *Left:* the real (solid line) and imaginary (dashed line) parts of $\chi^{(3)}_{THG}(\omega)$; *right:* the modulus of $\chi^{(3)}_{THG}(\omega)$

$\varepsilon = 0$ the imaginary part was vanishing. It means that the nonlinear resonant behaviour interesting from the practical point of view is accompanied by absorption.

For $z - 1/3 \gtrsim \alpha$ we obtain

$$\chi^{(3)}_{THG} = Bz^{-8}\left\{\frac{\pi\alpha}{6^{5/2}(z-1/3)^{3/2}} - a + \frac{c(3z-1)}{3} - \frac{i\pi}{6^{3/2}(z-1/3)^{1/2}}\right\}, \quad (21)$$

where $c = b - 24a - 4$.

For $\alpha = 0$ (21) also transforms into the asymptotic for the problem of zero damping. Again the peak values of the real and imaginary parts of $\chi^{(3)}$ are proportional to $\alpha^{-1/2}$. We note that even without the relaxation the real part of $\chi^{(3)}$ for $z \to 1/3 + 0$ does not diverge but reaches a constant negative value $-B_1 3^8 a$.

The $z = 1/2$ resonance produces no divergence in $\chi^{(3)}$. For $z - 1/2 \ll 1$ we have

$$\chi^{(3)}_{THG} = B2^8\left\{\frac{4}{5^{1/2}}\ln\left(\frac{3+5^{1/2}}{2}\right) 8 \cdot 3^{1/2}\arcsin\left(\frac{1}{2}\right)\right.$$
$$+ i\frac{2\pi}{5^{1/2}} + (2z-1)\left[10\cdot 4 - 88\cdot 3^{1/2}\arcsin\left(\frac{1}{2}\right)\right. \quad (22)$$
$$\left.\left.- \frac{32}{5^{1/2}}\ln\left(\frac{3+5^{1/2}}{2}\right) - i\frac{68\pi}{5^{3/2}}\right] + d\right\}.$$

Here d incorporates the contribution from the $z = 1/2$ resonance. For the case of zero damping it produces a small peak

$$d(z) = -4\cdot 2^{1/2}\pi\sqrt{1-2z}, \quad 0 \leq \frac{1}{2} - z \ll 1, \quad (23)$$

$$d(z) = -2(2z-1)\left(1 + \frac{i\pi}{2^{3/2}\sqrt{2z-1}}\right), \quad 0 \leq z - \frac{1}{2} \ll 1. \quad (24)$$

If we take relaxation into account d takes the form

$$d(z) = 8\pi\alpha^{1/2}(y - iA)\frac{\sqrt{\sqrt{1+y^2} - y} - i\sqrt{\sqrt{1+y^2} + y}}{\sqrt{1+y^2}}, \qquad (25)$$

$$y = \frac{2z-1}{2\alpha} \lesssim 1,$$

Here A is a constant that emerges in the coefficient accompanying $f(2z)$ due to damping within the term proportional to α. As we ignore small corrections in the nominator throughout, the value of A is unknown.

Equation (25) slightly shifts the $z = 1/2$ peak. The magnitude of the peak is now proportional to $\alpha^{1/2}$. So relaxation does not change the $z = 1/2$ resonance qualitatively.

The curves of the real and imaginary parts of $\chi^{(3)}$ are presented in Fig. 1,a. The fine lines in the vicinity of resonances show square root divergencies for the case $\varepsilon = 0$ while bold curves correspond to nonzero damping. Fig. 1,b shows the frequency dependence of the modulus of $\chi^{(3)}$.

The authors thank Prof. V.Ya. Krivnov for discussions.

References

1. *Nonlinear Optics of Organics and Semiconductors*, T. Kobayashi (Ed.), (Springer, Berlin 1989)
2. *Nonlinear Optical Effects in Organic Polymers* J. Messier, F. Kajzar, P. Prasad and D. Ulrich (Eds.), (Kluwer, Dordrecht 1989)
3. *Nonlinear Optical and Electroactive Polymers* P.N. Prasad and D.R. Ulrich (Eds.), (Plenum, N.Y. 1988)
4. P.A. Chollet, F. Kajzar, J. Messier: Synth. Met. **18** 459 (1987)
5. M. Sinclair, D. McBranch, D. Moses, A.J. Heeger: Synth. Met. **28** D645 (1989)
6. T. Sugiyama, T. Wada, H. Sasabe: Synth. Met. **28** C323 (1989)
7. F. Kajzar, S. Etemad, G.L. Baker, J. Messier: Synth. Met. **17** 563 (1987)
8. E. Wintner, F. Krausz, G. Leising: Synth. Met. **28** D155 (1989)
9. A.J. Heeger, S. Kivelson, J.R. Schrieffer, W.P. Su: Rev. Mod. Phys. **60** 781 (1988)
10. S.A. Brazovskii, N.N. Kirova: Sov. Sci. Rev. A **5** 99 (1984)
11. W. Wu: Phys. Rev. Lett. **61** 1119 (1988)
12. D.M. Mackie, R.J. Cohen, A.J. Glick: Phys. Rev. B **39** 3442 (1989)
13. L.V.Keldysh: Zh. Eksp. Teor. Fiz. **47** 1515 (1964) [Sov. Phys. JETP **20** 1018 (1965)]

Nonlinear Optical Properties of $(A-B)_x$–Polymers

Yuri I. Dakhnovskii[1] and Andre D. Bandrauk[2]

[1] Institute of Chemical Physics, Kosygina St. 4, 117 977 Moscow, USSR
[2] Department of Chemistry, Sherbrooke University, Sherbrooke, Quebec, Canada

Nonlinear optical properties of $(A-B)_x$–polymers are studied in the noninteracting electron approximation. Using a Keldysh Green function method there was calculated the nonlinear susceptibility of the third harmonic generation as well as nonlinear index of refraction. It was shown that the second order susceptibilities are equal to zero.

1. Introduction

Recently nonlinear optical properties of conjugated polymers are of interest [1–4] of many investigators. These polymers give us unusually large nonlinear optical susceptibilities. The value of $\chi^{(3)}$ of the third harmonic generation for polyacetylene and polydiacetylene is approximately equal to 10^{-9} e.s.u. Another candidate for such optically active polymer is $(A-B)_x$-one. This linearly conjugated diatomic polymer does not possess the inversion symmetry and therefore it is possible to observe $\chi^{(2)}$, the nonlinear optical susceptibility of the second order (Fig. 1). The examples of such a polymer are polycarbonitrile, $(CH=N)_x$ or charge-transfer mixed stack compounds, where donor and acceptor molecules alternate along the stacking direction, tetrathiafulvalene-chloranil [5–6].

In this paper nonlinear optical properties of linearly–conjugated diatomic polymer will be studied in the continuum model using Keldysh Green's function technique [7]. We will calculate $\chi^{(2)}(-2\omega;\omega,\omega)$ and $\chi^{(3)}$ for the third harmonic generation and intensity–dependent index of refraction (IDIR). At first we will derive the continuum Hamiltonian of the $(A-B)_x$–polymer in an electric field and obtain the current operator which does not coincide with the usual one.

2. Continuum Hamiltonian

Our polymer can be described by the usual site Hamiltonian

$$H = \alpha \sum_n (-1)^n a_n^+ a_n + \beta \sum_n (a_{n+1}^+ a_n + a_n^+ a_{n+1}) \\ + \Delta_0 \sum_n (a_{n+1}^+ a_n + a_n^+ a_{n+1}). \quad (1)$$

The first term describes the non equivalence of the even and odd sites. The second term is a usual hopping one, the third one is due to the Peierls distortion. Let us transform this Hamiltonian into the k-space with the help of the transformation

$$a_n = N^{-1/2} \sum_n e^{ikna} a_k. \quad (2)$$

The Hamiltonian takes the following form

$$H = \alpha \sum_k a_{k+\pi}^+ a_k + 2\beta \sum_k \cos(ka) a_k^+ a_k \\ + i\Delta \sum_k \sin(ka) a_{k+\pi}^+ a_k. \quad (3)$$

If we introduce the k-variable in the extended zone, where $-\pi/a \leq k \leq \pi/a$ and separate the summation into positive and negative ranges of k, we obtain

$$H = \alpha \sum_k (c_k^+ b_k + b_k^+ c_k) + 2\beta \sum_k \sin(ka)(b_k^+ b_k - c^+ c_k) \\ + i\Delta \sum_k \cos(ka)(c_k^+ b_k - b_k^+ c_k), \quad (4)$$

where we have introduced the following operators

$$c_k \equiv a_{k+\pi}; \quad c_k^+ \equiv a_{k+\pi}^+; \quad b_k \equiv a_{k-\pi}; \quad b_k^+ \equiv a_{k-\pi}^+. \quad (5)$$

Let us introduce the two–component fermionic field Ψ the Hamiltonian (4) can be written as

$$H = \sum \Psi_k^+ [2\beta \sin(ka)\hat{\sigma}_3 + \Delta \cos(ka)\hat{\sigma}_2 + \alpha\hat{\sigma}_1]\Psi_k, \quad (6)$$

where

$$\Psi = \begin{pmatrix} b_k \\ c_k \end{pmatrix} \quad \Psi_k^+ = \begin{pmatrix} b_k^+ \\ c_k^+ \end{pmatrix}.$$

$\hat{\sigma}_1, \hat{\sigma}_2, \hat{\sigma}_3$ are the Pauli matrices.

An electromagnetic field can be taken into account by a usual way, shifting the momentum k as following

$$k \Rightarrow k + e/cA,$$

where A is a vector potential of the electromagnetic field. After this transformation the Hamiltonian has the form

$$H = \sum_k \Psi_k^+ (2\beta \sin(k + e/cA) a \hat{\sigma}_3 \qquad (7)$$
$$+ i\Delta_0 \cos(k + e/cA) a \hat{\sigma}_2 + \alpha \hat{\sigma}_1) \Psi_k .$$

In a weak electromagnetic field equation (7) has the following form

$$H = H_0 + H_1 , \qquad (8)$$

where

$$H_0 = \sum_k \Psi_k^+ (2\beta \sin(ka) \hat{\sigma}_3 + \Delta_0 \cos(ka) \hat{\sigma}_2 + \alpha \hat{\sigma}_1) \Psi_k ; \qquad (9a)$$

$$H_1 = \frac{ea}{c} \sum_k \Psi_k^+ (2\beta \cos(ka) \hat{\sigma}_3 - \Delta_0 \sin(ka) \hat{\sigma}_2) \Psi_k A . \qquad (9b)$$

The usual way to introduce the continuum Hamiltonian is to expand all the functions and operators over k up to the linear terms. i.e., $ka \ll 1$. Then the continuum Hamiltonian has the following form

$$H_0 = a \sum \Psi_k^+ (2\beta k \hat{\sigma}_3 + \Delta_0 \hat{\sigma}_2 + \alpha \hat{\sigma}_1) \Psi_k ; \qquad (10a)$$

$$H_1 = \frac{ea^2}{c} \sum_k \Psi_k^+ (2\beta \hat{\sigma}_3 - \Delta_0 k a \hat{\sigma}_2) \Psi_k A . \qquad (10b)$$

Usually the second term in expression (10b) is small due to the inequality $\Delta_0/2\beta \ll 1$, but in our case for the calculation of $\chi^{(2)}$ it is of importance. Besides this term $\chi^{(2)}$ is equal to zero. From the Hamiltonian H_1 the current operator has the following form

$$I_k = -ea^2 \Psi_k^+ (2\hbar v_F \hat{\sigma}_3 - \Delta_0 k a \hat{\sigma}_2) \Psi_k . \qquad (11)$$

This operator will be used for the calculation of the nonlinear susceptibilities.

3. Nonlinear Optical Susceptibility

Because of the violence of the inversion symmetry we hope that there is a second order susceptibility in (A-B) polymers. For calculations we have used the Keldysh Green's function technique. Usually nonlinear optical susceptibilities are defined from the expression of the polarization vector over applied electric fields

$$P = \chi^{(1)} E + \chi^{(2)} E \cdot E + \chi^{(3)} E \cdot E \cdot E ; \qquad (12)$$

here P is a polarization vector and E is applied electric field. But

$$\frac{dP}{dt} = \langle I(t) \rangle .$$ (13)

Hence it is necessary to calculate the average value of the current operator. Here $\langle \ldots \rangle$ means thermodynamic average over the initial states. Using (12, 13) and Keldysh Green's function method we can obtain the expression for nonlinear susceptibilities [3] of second order. As it was noted in [7] the Keldysh nonequilibrium technique and Feynman diagrams are similar to those in equilibrium state, it simplifies the calculation.

$$\chi^{(2)}(-2\omega;\omega,\omega) = \frac{e^3}{2\omega^3\hbar^2} \int \frac{dk\,d\omega_1}{(2\pi)^2} \text{Tr} \left\{ \widehat{I}_k \widehat{G}(\omega_1) \right.$$
$$\left. \times \widehat{\tau}_3 \widehat{I}_k \widehat{G}(\omega_1 + \omega) \widehat{\tau}_3 \widehat{I}_k \widehat{G}(\omega_1 + 2\omega) \right\}_{21} .$$ (14)

Here $\widehat{I}_k$ is defined by equation (11), $\widehat{G}$ is a Green's function matrix in the Keldysh space and in two-component fermionic one; $\widehat{\tau}_3$ is a Pauli matrix in the Keldysh space. After averaging over the initial states we must take the matrix element (21) in the Keldysh space.

The components of the Keldysh Green's function space can be easily calculated according to [7]

$$G_{11} = \frac{\omega + v_F k \widehat{\sigma}_3 - \Delta_0 \widehat{\sigma}_2 + \alpha \widehat{\sigma}_1}{\omega^2 - \omega_k^2 + i\eta} ;$$

$$G_{22} = -\frac{\omega + v_F k \widehat{\sigma}_3 - \Delta_0 \widehat{\sigma}_2 + \alpha \widehat{\sigma}_1}{\omega^2 - \omega_k^2 - i\eta} ;$$ (15)

$$G_{21} = -\frac{i\pi}{\omega_k}[\omega_k + v_F k \widehat{\sigma}_3 - \Delta_0 \widehat{\sigma}_2 + \alpha \widehat{\sigma}_1]\delta(\omega - \omega_k) ;$$

$$G_{12} = \frac{i\pi}{\omega_k}[\omega_k - v_F k \widehat{\sigma}_3 + \Delta_0 \widehat{\sigma}_2 - \alpha \widehat{\sigma}_1]\delta(\omega + \omega_k) ,$$

where

$$\omega_k^2 = \alpha^2 + \Delta_0^2 + \hbar^2 v_F^2 k^2 ; \quad 2\beta a = v_F .$$

But after the calculation $\chi^{(2)}(-2\omega;\omega,\omega)$ is vanishing. Therefore it is necessary to calculate the susceptibilities of the third order. For $\chi^{(3)}(-3\omega;\omega,\omega,\omega)$ similar way

$$\chi^{(3)}(-3\omega;\omega,\omega,\omega) = \frac{B_1}{z^8} \{3(1-8z^2)f(3z) - 8(1-4z^2)f(2z) + (5-8z^2)f(z)\} ;$$ (16)

with

$$B_1 = e^4(\hbar v_F)^3 a / 1152\pi \Delta_0^6 \nu ; \quad z \equiv \hbar\omega/2\Delta_0 ;$$

and

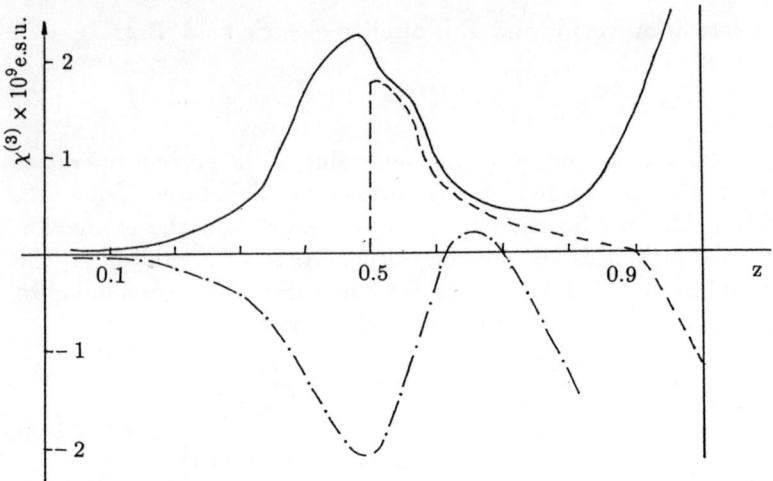

Fig. 1. The calculated $\chi^{(3)}_{\text{THG}}(\omega)$. The dot–dashed line is the real part; the dashed line is the imaginary part; and the solid line is the absolute value

$$f(z) \equiv \begin{cases} \frac{\arcsin(z)}{z(1-z^2)^{1/2}} \; ; & z \leq 1 \; , \\ \frac{1}{z(1-z^2)^{1/2}} \left(-\frac{i\pi}{2} - \cosh^{-1}(z) \right) \; ; & z \geq 1 \; , \end{cases} \quad (17)$$

this function coincides with that of calculated in ref. [3] for polyacetylene but $\Delta_0^2 = (\Delta^2 + \alpha^2)$.

Next we will calculate the value of intensity dependent index of refraction $\chi^{(3)}(\omega)$ defined as in ref. [3]

$$\chi^{(3)}_{\text{IDIR}}(\omega) \equiv \frac{1}{3}[\chi^{(3)}(-\omega;-\omega,\omega,\omega) \\ + \chi^{(3)}(-\omega;\omega,-\omega,\omega) + \chi^{(3)}(-\omega;\omega,\omega,\omega)] \; . \quad (18)$$

This function may be obtained in a similar way using Keldysh Green's function technique. After the tedious calculations we have

$$\chi^{(3)}_{\text{IDIR}}(\omega) = -\frac{B_2}{z^8}\left[(1-4z^2)f(2z) + \frac{(1-4z^2)}{2(1-z^2)^2} \right. \\ \left. - \frac{12z^4 - 7z^2 + 2}{2(1-z^2)^2} f(z) - g(z) \right] \; , \quad (19)$$

where

$$B_2 = e^4(\hbar v_F)^3 a/48\pi\Delta_0^6 \nu$$

and

$$g(z) = -4.6(6)z^4 - 14.054918z^6 - 26.775974z^8 \; . \quad (20)$$

This function is a regularization one according to Pauli-Villars procedure [9]. This regularization is due to divergence of $\chi^{(3)}_{\text{IDIR}}(\omega)$ at $\omega \Rightarrow 0$. Equation (19) is different to that of obtained in ref. [3]. As in the case of polyacetylene there is no infinite absorption at $2z = 1$, two photon resonance is suppressed by the factor $(1-4z^2)$.

The intensity dependent index of refraction is of the same order of magnitude as $\chi^{(3)}_{THG}$. For practical purposes it is of importance that the ratio of Im $\chi^{(3)}_{IDIR}$/Re$\chi^{(3)}_{IDIR}$ must be as small as possible. It is a necessary condition for waveguides and it is valid in wide range of $z(\hbar\omega/2\Delta_0)$. The frequency dependence $\chi^{(3)}_{IDIR}$ is shown on Fig. . The main difference from that of obtained in ref. [3] is that the real part of $\chi^{(3)}_{IDIR}(\omega)$ has both minimum at $2\hbar\omega = 2\Delta_0$ and maximum at $\hbar\omega/2\Delta = 0.63$. Therefore there is another ω-dependence of $|\chi^{(3)}_{IDIR}(\omega)|$.

4. Conclusions

Nonlinear optical properties of $(A - B)_x$-polymer have been considered. We have calculated susceptibility of the second order as we hoped to have a nonvanishing value. We have derived the Hamiltonian and the current operator which differs from that of obtained earlier. I was shown that the second order susceptibility is equal to zero. Therefore we have calculated the third order susceptibilities. It was found that the third harmonic generation and $\chi^{(3)}_{THG}(-3\omega;\omega,\omega,\omega)$ coincides with that of obtained by Wu [3]. But the value of the intensity dependent index of refraction is different (Fig.). There is minimum at $\hbar\omega = \Delta_0(z = 0.5)$ and maximum at $z = 0.63$.

One of the authors (Yu.D.) thanks A.A. Ovchinnikov, S.A. Brazovskii, V.Ya. Krivnov and N.N. Kirova for useful discussions.

References

1. F. Kajzar, S. Etemad, G.L. Baker and J. Messier: Synth. Met. **17** 563 (1987)
2. A.J. Heeger, D. Moses and M. Sinclair: Synth. Met. **17** 343 (1987)
3. W. Wu, Phys. Rev. Lett. **61** 1119 (1989)
4. D.M. Mackie, R.J. Cohen and A.J. Glick: Phys. Rev. **B39** 3442 (1989)
5. H.J. Rice and E.J. Mele: Phys. Rev. Lett. **49** 1455 (1982)
6. T. Martin and D.K. Campbell: Phys. Rev. **B35** 7732 (1987)
7. L.V. Keldysh: Zh. Eksp. Teor. Fiz. **47** 515 (1964) [Sov. Phys. JETP **20** 1018 (1965)]
8. Y.R. Shen: *Principles of Nonlinear Optics* (Whiley, New York 1987)
9. C. Itzykson and J.B. Zuber, *Quantum Field Theory* (McGraw-Hill, New York 1980)

Application of the Method of Cyclic Permutations to the Calculation of Many–Electron Systems. Polaron States in Emery Model

V.Ya. Krivnov[1], *A.A. Ovchinnikov*[1] *and V.O. Cheranovskii*[2]

[1] Institute of Chemical Physics, Kosygina St. 4, 117 977 Moscow, USSR
[2] Kharkov State University, Institute of Chemistry, Kharkov, USSR

The problem of the reducing of some strong correlated models to the spin ones is discussed. The Hubbard model with infinite interaction on rectangular lattice and one–dimensional Emery model are considered. For latter the spectrum of polaron states, which are formed by an additional hole, is calculated.

The problem of highly correlated electrons is the central point in the theory of high-T_c superconductors [1]. The simplest model of the strong correlated systems is the Hubbard model. For large on site repulsion at half-filling it can be reduced to the Heisenberg spin model. But for the case of not half-filling a similar reducing is possible in 1d case only.

The purpose of this paper is to show that more complicated models like one–band Hubbard one on rectangular lattice with $U = \infty$ and Emery model can be described in terms of spinless fermions and the operators of cyclic spin permutations. Such a factorized representation in certain cases allows to reduce an initial model to a spin one. In consequence it is possible to examine the spectrum of such spin Hamiltonians using different methods of theory of spin systems.

Let us consider a rectangular lattice described by Hubbard Hamiltonian with infinite repulsion and for arbitrary electron densities. The finding of the spectrum of such kind of the lattices is a complicated kinematic problem which exact solution is only known for 1d case. The wave function of this system can be written as a superposition

$$\Psi = \sum \Phi(n_1, n_2, \ldots, n_s) \Theta(\sigma_1, \sigma_2, \ldots, \sigma_s) , \qquad (1)$$

where Φ is a spatial part of wave function; n_j are the numbers of the occupied sites; Θ is a function of spin variables describing the spin configuration of a lattice. These processes lead to the two new configurations (Fig. 1). For the first one the numeration of electrons is not changed and the number of the first occupied site is enlarged to one. But for second configuration the numbers of electrons have been already changed: the first, the second and the third electrons should be the third, the first and the second one respectively, i.e. to keep the successive numeration of electrons in the lattice row it is necessary to make the cyclic permutation of numbers of electrons which are situated between the first and the fifth sites. Apart from the evident change of coordinate function this leads to cyclic permutation of the first three spin variables in the spin function Θ, i.e., the spin configuration of the lattice is changed.

Let us numerate the electrons and sites successively along the lattice rows and consider 5-electron configuration. The action of H on this configuration transfers the electrons to the neighboring unoccupied sites.

The corresponding permutation of spins can be written in standard form

$$Q_{13} = \begin{pmatrix} 1 & 2 & 3 \\ 3 & 1 & 2 \end{pmatrix},$$

here the first row is an initial spin configuration and the second one is the configuration after cyclic permutation. (We note that the invariance of Q with respect to the electron transfer along the lattice is connected with a spin degeneracy of the spectrum of a Hubbard chain with infinite interaction). In general case the action of H on any electron configuration can be considered in analogy with the preceding. It is easy to see that the Hubbard Hamiltonian of model under consideration has a following form

$$H = \sum_{n<m} c_n^+ c_m Q_{kl}^+ + c_m^+ c_n Q_{kl} , \qquad (2)$$

where c_n are operators of spinless fermions; Q_{kl} are cyclic permutations of spin variables for the electrons locating between i-th and j-th sites.

This representation allows to construct the rather simple algorithms for exact small clusters calculations, which are more succinct in comparison with well known algorithm of Takahashi [2].

Especially the method of cyclic permutations is suitable to the strong anisotropic systems. Such systems have been considered in our work [3].

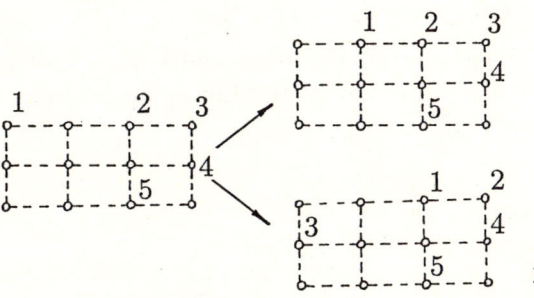

Fig. 1.

They are strips consisting of n-site segments and are described by Hubbard Hamiltonian with $U = \infty$ and two transfer integrals t and $t_\perp$ ($\alpha = |t_\perp/t| \ll 1$). The perturbation theory (PT) in α takes into account the electron transfers between the segments. These transfers will lead to cyclic spin permutations if electrons are successively numerated along segments beginning from the upper one.

If each segment contains the same number of electrons the first order of PT gives zero. Then using a method of cyclic permutations and summing on lattice variables in second PT order we obtain the spin Hamiltonian

$$H = \sum H_{ii+1},$$

$$H_{ii+1} \sum_{kl=1}^{s} \sum_{pq=1}^{s+1} J(klpq)\{Q_{kl}^+(s-1)Q_{pq}(s-1) + Q_{lk}(s)Q_{qp}^+(s)\},$$ (3)

where H_{ii+1} is the Hamiltonian for segments i and $i+1$, J are effective exchange integrals, defined in terms coefficients of wave functions of isolated segments.

$$Q_{kl} = \begin{pmatrix} k & k+1 & \ldots & s+1 \\ s+1 & k & \ldots & s+l-1 \end{pmatrix}.$$

If segments contain different number of electrons spin degeneracy is resolved in first PT order. In this case an interaction Hamiltonian of two neighboring segments is

$$H = \sum_{kl} J(kl)\{Q_{kl}^+ c_i^+ c_{i+1} + Q_{kl} c_{i+1}^+ c_i\},$$ (4)

where c_i^+ is the creation operator of the state with $s+1$ electrons on i-th segment.

Another example of the strong correlation model, for which can be applied developed method, is Emery model [4]. It is proposed that this model is the most realistic one for the copper oxides. We will consider its simplified version corresponding to the following choice of the parameters (in standard notation):

$$U_d = \infty, \quad U_p = 0, \quad \varepsilon = \varepsilon_p - \varepsilon_d \gg |t_{pd}|.$$

For these parameters there is one hole per Cu atom and doping creates holes on O sites. In this case, the Hamiltonian of the Emery model can be transformed into following form [5–8]

$$H = N_h J + J/2 \sum_{n,\sigma,\sigma'}^{N} \sum_{j,j'=\pm 1} (\delta_{\sigma\sigma'}\delta_{j,-j'} + 4S_{2n}S_{\sigma\sigma'})a_{2n+j\sigma}^+ a_{an+j'\sigma'},$$ (5)

where N_h is a number of holes, $a_{2n+1\sigma}^+$ creates holes on O sites; S_{2n} is a copper spin operator; $S_{\sigma\sigma'}$ are matrix elements of spin 1/2 operator; Cu

(O) atoms are located at even (odd) sites; N is a number of Cu or O atoms; $J = t_{pd}^2 \varepsilon$. Besides, in (5) we have neglected the terms of order $t_{pd}^4 \varepsilon^3$, which lead to the superexchange interaction between the copper spins.

Although Hamiltonian (5) is an oversimplified version of initial Emery model, one believes that (5) keeps its main features. One of the most interesting questions relating to the model (5) is a problem of polaron formed by an additional hole, i.e. bound state of the hole moving along oxygen atoms and interacting with copper spins. It was shown in [5, 7] that unlike Hubbard model the polaron is, most likely, nonmagnetic. Unfortunately, the polaron problem is not solved exactly and the detail information about its properties is missed. Therefore the numerical calculations of finite systems are of interest. We will consider the one-dimensional case keeping in mind that the spectrum of 1d qualitatively looks like that of in 2d [5].

Applying the method of cyclic permutations to the model (5) with one oxygen hole we can reduce it to a spin one and, besides, the total momentum is a parameter of a spin Hamiltonian

$$H = J \left[\sum c_i^+ c_i P_{ii\pm 1} + \sum (c_i^+ c_{i+1} + c_{i+1}^+ c_i) + (c_1^+ c_{N+1} Q^+ + c_{N+1}^+ c_1 Q) \right], \quad (6)$$

where c_i is spinless fermion operator defined by (4); P and Q are cyclic spin permutations of two and n+1 spins respectively.

Averaging over wave functions corresponding to a fixed value of the total momentum k and summing on lattice variables we we can written Hamiltonian (6) in a simpler form

$$H = J(P_{12} + P_{1N+1} + \exp(ik)Q + \exp(-ik)Q^+). \quad (7)$$

The use of this representation allows to study the states of an arbitrary multiplicity and reduces the dimension of the basis. This fact essentially simplifies the numerical calculations of the spectrum of (5).

Let us now discuss the results of the calculations of the finite clusters. (We considered the rings containing up to 28 sites).

In the Table 1 the ground state energy E_0 of rings with different values of N is shown.

Table 1.

N	6	7	8	9	10	11	12	14
E_0/J	−2.809	−2.848	−2.772	−2.902	−2.828	−2.879	−2.839	−2.856

As can be seen from Table 1 the dependence E_0 on N is very irregular. On the contrary, $E_0(N, S)$ is a monotonic, fast convergent function of N at fixed value of total spin S and values of $E_0(\infty, m)$ ($m = S_{\max} - S$) are

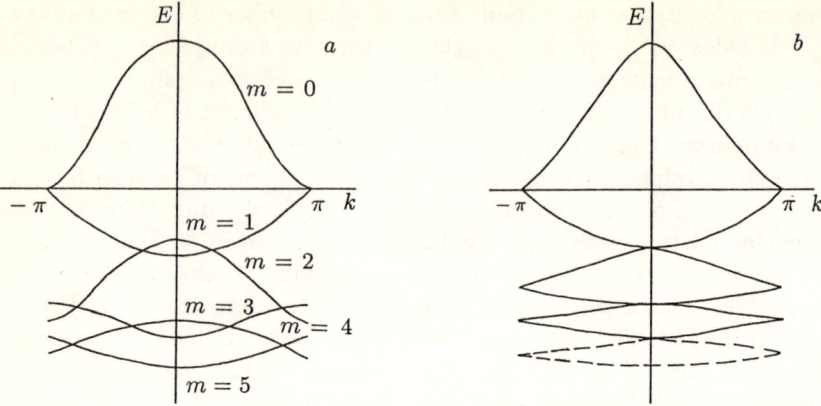

Fig. 2. k-Dependence of the state with lowest energy for different values of a total spin $S = S_{\max} - m$ (m is number of upturned spins): a) for $N = 10$, b) for $N = \infty$ (according to [5])

$E(\infty, 1) = -2.472\,J$, $E(\infty, 2) = -2.697\,J$, $E(\infty, 3) = -2.778\,J$, $E(\infty, 4) = -2.818\,J$, $E(\infty, 5) = -2.841J\,J$.

The momentum dependence of the lowest state energy for different values of the total spin S is shown in Fig. 2

We note that the ferromagnetic state $(S = S_{\max})$ is an usual band state of the hole. The bound state arises at $S \leq S\max - 1$. The energy of the state with $S = S_{\max} - 1$ (a one upturned spin) can be found exactly for $N \to \infty$. In particular, for this state the energy at $k = 0$ equals $-2.472\,J$.

As can be seen from Fig. 2 and Table 2 the energy decreases with decreasing S and the ground state is, most likely, a singlet when $N \to \infty$, i.e. a polaron is a nonmagnetic one.

Table 2. The energies E_m/J for $N = 8$ and 10, $k = 0$ and $k = \pi$

	m	1	2	3	4	5
$N = 8$	$k = 0$	-2.472	-2.244	-2.772	-2.724	—
	$k = \pi$	0	-2.687	-2.611	-2.763	—
$N = 10$	$k = 0$	-2.472	-2.333	-2.779	-2.774	-2.822
	$k = \pi$	0	-2.695	-2.664	-2.806	-2.785

The overlap at $k = 0$ or $k = \pi$ between the bands with the two successive values of m decreases with increasing N (when $k = 0$ and $k = \pi$ the energies E_m/J for $N = 8$ and 10 are given in Table 2). This fact confirms a hypothesis of Glasman and Ioselevich [5] about neighboring bands being touched with each other at $k = 0$ or $k = \pi$ (according to [5] the forms of spectra in one and two dimensions qualitatively coincide).

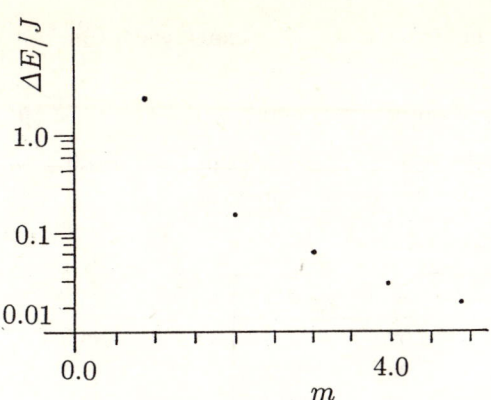

Fig. 3.

Another important feature of the spectrum is sharp decreasing of the bandwidth $\Delta E(m)$ ($\Delta E(m) = |E_m(\pi) - E_m(0)|$) with increasing m. The dependence $\ln(\Delta E/J)$ on m is shown in Fig. 3.

This dependence and the form of the spectrum in Fig. 2 allows to suppose that the band with $S = 0$ is dispersionless when $N \to \infty$. Thus the effective polaron mass equals infinity: *the nonmagnetic polaron will be localized* (at least, as long as the direct hopping integral $t_{pp} = 0$).

An important question which arises is the question about a spin structure of the polaron. To determine it we have calculated spin correlation functions of different types. The correlators G_1 and G_2 define a spin structure of the polaron around the hole.

$$G_1 = 1/N \sum_n \langle S_{2n} S_{2n+2} c^+_{2n+1} c_{2n+1} \rangle ,$$
$$G_2(l) = 1/N \sum_n \langle S_{2n+1} S_{2n+2l} \rangle .$$
(8)

A correlation of the pair of copper spins which neighbor the hole (Table 3) is clearly ferromagnetic, i.e. the hole tends to set up its neighboring copper spins in parallels.

Table 3. The correlator G_1 in ground state for rings with different N

N	6	7	8	9	10	11
G_1	0.237	0.238	0.232	0.235	0.236	0.236

The correlations between the hole spin and copper spins defined by the function G_2 are antiferromagnetic. We note that the correlator G_2 for near O and Cu atoms weakly depend on the system total spin. As for a polaron size it depends on S. For $S = S_{\max} - 1$ and $N \to \infty$ the exact polaron wave function and all correlators can be found easily, besides, the correlators for

$N = 10$ practically coincide with the exact ones. The exact wave function of this state is

$$\varphi(n)(-1)^{(n+1)/2} \exp(-v|n|), \quad v = 0.722, \tag{9}$$

where n is a distance between a hole and an upturned spin. According to (9) and Table 4 the polaron size is about a lattice constant.

Table 4. Values of $G_2(l)$ for $N = 10$ with different m

$m\backslash l$	1	2	3	4	5
1	−0.474	0.048	−0.046	−0.022	−0.027
2	−0.485	0.147	−0.089	0.035	−0.023
3	−0.487	0.160	−0.088	0.038	−0.014
4	−0.487	0.170	−0.095	0.044	−0.012
5	−0.488	0.160	−0.080	0.059	−0.026

For states with two and more upturned spins the polaron size is considerably larger. At least it more than the size of the considered systems. For its exact determination the calculations of larger systems are needed.

The correlator G_3 is copper spin correlation function averaged over the hole motion

$$G_3(l - n) = \langle S_{2n} S_{2l} \rangle. \tag{10}$$

Table 5. Values of $G_3(n)$ for $N = 10$, $m = 5$

n	1	2	3	4	5
G_3	−0.215	−0.045	−0.042	−0.040	0.084

It is seen from Table (4) that the motion of a hole induces antiferromagnetic correlations for the neighboring copper spins (unlike the correlators G_2) but the correlations between more distant spins are not antiferromagnetic and spin structure of copper spin subsystem is frustrated.

References

1. P.W. Anderson: Science **235** 1196 (1987)
2. M. Takahashi: J. Phys. Soc. Japan **51** 3475 (1982)
3. V.Ya. Krivnov, A.A. Ovchinnicov, V.O. Cheranovskii: in this book
4. V.J. Emery: Phys. Rev. Lett. **58** 2794 (1987)
5. L.I. Glasman, A.S. Ioselevich: Pis'ma Zh. Eksp. Teor. Fiz. **47** 464 (1988)
6. A.F. Barabanov, L.A. Maksimov, G.V. Uimin: Zh. Eksp. Teor. Fis. **96** 655 (1989)
7. M.W. Long: J. Phys.: Cond. Matter **1** 9421 (1989)
8. F.C. Zhang and T.M. Rice: Phys. Rev. B **37** 3759 (1988)

From Incomplete Allowance for Electron Correlation to the Full CI in π-Systems. The Variational Operator Approach

A.V. Luzanov, Yu.F. Peash and V.V. Ivanov

Kharkov State University, Department of Chemistry, 310 077 Kharkov, USSR

1. Introduction

Electron correlation in the real physical systems may be investigated in many sophisticated ways and means, and from among them one ought to prefer such a general approach that leads to approximate but reliable consideration as well as to the exact one. Therefore, when describing the electronic structure of molecular systems we favour the most extended *modus operandi* which is supplied us by the operator algebra. An important advantage of the study in terms of the operators is its ability to set up the electronic model in invariant (more precisely, covariant) terms. It means, the main equations have the same form no matter how the orbital frame is constructed.

Adhering to this covariant strategy [1], we shall derive here some matrix equations related to the full configuration interaction (FCI) which correspond with the exact solution of Shrödinger equation in a finite-dimensional basis. Next, performing the FCI calculation for the electronic systems of some interest, one gives a possibility to juxtapose the exact results and that of the well–known or certain novel approximate models.

2. FCI as Extension of Geminals Problem

Now we consider a two-electron system with the proper Schrödinger equation for $\Psi = \Psi(r, r')$

$$\widehat{h}(r)\Psi + \widehat{h}(r')\Psi + \Psi/|r - r'| = E\Psi \tag{1}$$

and translate it into the operator equation. For this, regard the two-electron wave function (geminal) $\Psi(r, r')$ as a kernel of some integral operator X. In so doing, the first member in the left-side hand of (1) is hX with h as the integral operator of $\widehat{h}(r)$, the second is Xh and the last term represents, in fact, the exchange operator $\widehat{K}(X)$ in the such sense that

$$\widehat{K}(X) = \text{tr}_{(2)} P_{12}^0 g(12) X(2), \tag{2}$$

where $g(12)$ denotes the two-particle operator of electron repulsion operator, P_{12}^0 is the spinless transposition, and tr is taking operator trace. Thence, the 2-electron equation (1) is equivalent to 1-electron operator (or matrix) equation

$$hX + Xh + \widehat{K}(X) = EX .\tag{3}$$

Extending this reasoning, treat a spinless $2n$-electron wave function $|\overset{\circ}{\Psi}\rangle$ as the kernel of a certain spinless operator say $\overset{\circ}{X}_n = \overset{\circ}{X}(1,\dots n)$:

$$\overset{\circ}{\Psi} = \Psi(r_1,\dots,r_n,r'_1,\dots,r'_n) \longrightarrow \overset{\circ}{X}_n .\tag{4}$$

In (4) one supposes the unprimed variables referred to electrons with up spin ket $|\alpha\rangle$ and the primed variables to electrons with down spin ket $|\beta\rangle$, so that for total wave function there is the presentation by the antisymmetrized product ($N = 2n$)

$$|\Psi\rangle = A_N |\overset{\circ}{\Psi}\rangle |\alpha(1)\dots\alpha(n)\rangle |\beta(n+1)\dots\beta(N)\rangle .\tag{5}$$

By dint of applying the direct operatorial methods [1, 2] one can establish the N-electron Schrödinger equation for (5) as being in one-to-one correspondence with the next operator equation generalizing (3)

$$H_n \overset{\circ}{X}_n + \overset{\circ}{X}_n H_n + \widehat{K}(\overset{\circ}{X}_n) = E \overset{\circ}{X}_n \tag{6}$$

with H_n having the usual meaning of n-electron Hamiltonian and the exchange operator $\widehat{K}$ extending (2) follows

$$\widehat{K}(\overset{\circ}{X}) = n \overset{\circ}{A}_n [\text{tr}_{(n+1)} P_{1,n+1}^0 g(1,n+1) \overset{\circ}{X}(2\dots n+1)] \overset{\circ}{A}_n ,\tag{7}$$

where $\overset{\circ}{A}_n$ is the spinless n-particle antisymmetrizer.

The stated equation (7) offers a convenient method for the numerical computation within the FCI. For this purpose one should chose a spinless orbital basis $\{\chi_i\}$ and view $\overset{\circ}{X}_n$ as the multiindex matrix

$$\overset{\circ}{X}_n = \| \overset{\circ}{X}_{IJ} \| , \quad X_{IJ} = X_{i_1\dots i_n; j_1\dots j_n} \tag{8}$$

in the basis adopted. Due to (5) the $|\overset{\circ}{\Psi}\rangle$ must be antisymmetric in the integers $1, 2, \dots, n$; thus

$$\overset{\circ}{X}_n = \overset{\circ}{A}_n \overset{\circ}{X}_n \overset{\circ}{A}_n ,\tag{9}$$

i.e. the matrix elements (8) antisymmetric in $i_1,\dots,i_n$, as in $j_1,\dots,j_n$. The given restriction does not cover all requirements for $\overset{\circ}{X}$. Yet, involving all such requirements adds to difficulties in a numerical work, and it is rather practical to ignore them allowing the computational process itself to bring down to the correct $\overset{\circ}{X}$. These features are somewhat similar to that of [3, 4], but the main distinction of our formulation is its matrix form (6) determining a vectorizability of the algorithm proposed.

Equally important is a possibility to produce the simplified electronic models directly from $\overset{\circ}{X}$. As example, assuming the factorization

$$\overset{\circ}{X}_n = \overset{\circ}{A}_n X(1)\ldots X(n) \qquad (10)$$

one gets the approximation which precisely coincides with the so called antisymmetrized geminal power state (AGP). Owing to that simple algebraic form (10), one gains not only a new insight into the AGP as a strict framework of the BCS model, but also the computationally suitable matrix implementation of the APG in which X is the variational one-electron operator.

3. FCI on Base of Spin Flipping Operators

The alternative approach to the covariant framework of the electronic models has been outlined in [6] by applying the idea of spin flipping to the upmost decoupled state $|\Phi_\uparrow\rangle$ with maximum spin (like a ferromagnetic state) to provide for the state $|\Psi\rangle$ with a given spin s. In a conventional orbital language this idea was more or less known [7]. Here again, "operatorial" mode of thinking offers considerable scope for the further developments.

In order to anticipate the desired equation for the FCI, once more examine the two-electron problem in a minimal (for simplicity) basis of two atomic orbitals $|\chi_1\rangle$, $|\chi_2\rangle$. As an initial state, the fully decoupled (triplet) state

$$|\Phi_\uparrow(12)\rangle = A_2|\chi_1(1)\chi_2(2)\rangle|\alpha(1)\alpha(2)\rangle \qquad (11)$$

is taken, so the singlet state is found to be the next transformation of (11)

$$|\Psi(12)\rangle = A_2 t(1) s_-(1)|\Phi_\uparrow(12)\rangle . \qquad (12)$$

In (14) the spinless one-electron operator t is variational one and goes along with spin flipping performed by lowering operator $s_- = |\beta\rangle\langle\alpha|$. It is readily to find that the appropriate Schrödinger equation converts to the one-electron expression

$$h^\beta t - t h^\alpha - \widehat{K}(t) = \lambda_\downarrow t , \qquad (13)$$

where

$$h^{(\alpha)} = h + \widehat{J}(I) - \widehat{K}(I) , h^{(\beta)} = h + \widehat{J}(I) , \widehat{J}(X) = \text{tr}_{(2)} g(12) X(2) , \qquad (14)$$

and the number

$$\lambda_\downarrow = E(\Psi) - E(\Phi_\downarrow) \qquad (15)$$

is the electron coupling energy which acts as eigenvalue associated with t as the eigenvector of the problem (13).

Putting this reasoning into a more general form, we may now assume again a size of orbital basis $\{\chi_i\}$ to be equal to number of electrons N. Then, starting with

$$|\Phi_\uparrow\rangle = A_N |\chi_1(1)\ldots\chi_N(N)\rangle|\alpha(1)\ldots\alpha(N)\rangle \qquad (16)$$

as the state for maximum spin $N/2$, we produce a state $|\Psi\rangle$ for a spin $s < N/2$ by the way like (12)

$$|\Psi\rangle = A_N \overset{\circ}{T}(1\ldots k) s_-(1)\ldots s_-(k)|\Phi_\uparrow\rangle . \qquad (17)$$

In the above construction the spinless k-particle operator $\overset{\circ}{T} = \overset{\circ}{T}(1\ldots k)$ involves the whole information concerning $|\Psi\rangle$. An integer k to be a number of spin flipping is determined by the relation

$$k = N/2 - s . \qquad (18)$$

By use of the technique [2], one can deduce the equation for $\overset{\circ}{T}$, taking the place of the Schrödinger equation for Ψ:

$$H_k^{(\beta)} \overset{\circ}{T} + \overset{\circ}{T} H_k^{(\alpha)} - \widehat{K}(\overset{\circ}{T}) = \lambda_\downarrow^{(k)} \overset{\circ}{T} , \qquad (19)$$

$H_k^{(\alpha)}$ and $H_k^{(\beta)}$ being the modified k-electron Hamiltonians in which h is replaced respectively by $-h^{(\alpha)}$ and $h^{(\beta)}$ (14). The 7value $\lambda_\downarrow^{(k)}$ is the corresponding coupling energy defined as (17).

The new eigenvalue problem (18) presents some other operatorial frame of the FCI and may be easily expanded into a case of arbitrary basis of $r \geq N$ orbitals. The interesting property of that frame is an unusual form in the condition for spin purity of (17), namely, one must impose on the $\overset{\circ}{T}$ the simple restriction as (9) and the special requirement to be traceless operator in the following sense

$$\text{tr}_{(k)} \overset{\circ}{T}(1\ldots k) = 0 . \qquad (20)$$

Admittedly, in the computational scheme for the FCI problem (19) it is too convenient to disregard such additional restriction in order to deal with a simple matrix algorithm for finding $\overset{\circ}{T}$. The latter is specifically fit for calculating open-shell systems with arbitrary value of spin.

The description in terms of $\overset{\circ}{T}$ naturally leads to deriving a number of pertinent approximations. Thus, letting

$$\overset{\circ}{T} = \overset{\circ}{A}_k t(1)\ldots t(k) \qquad (21)$$

one obtains the electronic model named in [6] as "one-particle amplitudes for spin-coupling" (OASC). This factorized pattern of $\overset{\circ}{T}$ is further abbreviated to OASC/1.

The OASC/1 has given a fairly good picture for the spin density distribution and other properties of π-radicals and π-triplets [8]. Nonetheless, the worth of this model decreases with increasing a size of π-system studied (e.g. for lengthy polyene chains). This drawback is related with a lack of the so called size-consistency [9] in the OASC/1 as well as in AGP and the extended Hartree-Fock method (EHF) also used for quasi-one-dimensional systems.

The model free of the such shortcoming is made of (21) by its slight modification

$$\overset{\circ}{T} = \overset{\circ}{A}_n \, t_1(1) \ldots t_a(a) \ldots t_n(n) \, \overset{\circ}{A}_n \, , \qquad (22)$$

where each operator t_a pertaining to the a th spin flipping may be now independent of all other t's. For numerical realization, this new model, called OASC/2, is far more difficult than OASC/1, but it comes to more reliable results in return.

4. Calculations on Small π-Systems

In this section the aforementioned models are studied for π-systems to search the correlation effects by itself and to expound the situation about relative merits and demerits of the models. The calculations have been performed at the following levels of a conventional π-approximation:

I) using nonzero resonance integrals $\beta_0 = -2.274$ eV and two-center integrals of electron repulsion, $\gamma_{\mu\nu}$, in the form due to Ohno with $\gamma_0 = \gamma_{\mu\mu} = 11.13$ eV;
II) using $\beta = -2.4$ eV and $\gamma_{\mu\nu}$ by Mataga approximation;
III) using the Hubbard model ($\gamma_{\mu\nu} = \gamma_0 \delta_{\mu\nu}$) determined by the ratio $U = \gamma_0/\beta_0$.

Below all energy values are also expressed in eV. Here we confine ourselves to the linear polyenes with N carbon atoms ($C_N H_{2N+2}$). All these π-systems were assumed to be planar, the value 120^0 was assigned to the C–C–C bond angle.

First, we examine efficiency of AGP, EHF and OASC in respect to the full CI. The corresponding data for the ground states with spin values $s = 0$ and $s = 1$ (the parametrization I) are listed in the Table 1.

Table 1. Fraction of correlation energy (as a percentage) obtained in the main models. Top: singlet state; bottom: triplet state

model\N	2	4	6	8	10
AGP	100	46	33	23	18
OASC/1	100	80	66	58	51
ENF	100	90	82	78	72
OASC/2	100	98	95	–	–

model\N	2	4	6	8	10
EHF	–	77	65	60	56
OASC/1	–	100	83	74	68
OASC/2	–	100	97	95	–

As seen from these figures, only OASC/2 make fairly steady allowance for correlation effects. (It is rather surprising to see a little efficiency of the AGP model which gives apparently negligible inclusion of correlation as molecular size (N) increases.) It is worthy to denote a high localization (about appropriate double bound) for each electron coupling matrix t_a associated with a proper spin flipping. From whence the k-particle $\overset{\circ}{T}$ in the form (22) automatically becomes nearly traceless to obey (20). Moreover, as N makes larger concurrently with spin s, the coupling energy $\lambda_\downarrow^{(k)}$ is proved to be a more additive in the sense that $\lambda_\downarrow^{(k)} \approx k\lambda_\downarrow$ for the small k/N:

Table 2.

	$N = 10$	$N = 15$	$N = 20$
$-\lambda_\downarrow$	5.63	5.69	5.71
$-\lambda_\downarrow^{(2)}$	10.82	11.20	11.72
$-\lambda_\downarrow^{(3)}$	15.09	16.12	16.73

This is found to be in evident correspondence with magnon theory of elementary excitations for high-spin systems.

Some attention may be called to a dependence of a "specific" energy correlation, $\varepsilon_{\text{corr}} = E_{\text{corr}}/N$, on a chain size N under different parameterization schemes mentioned above. The net result of the study is following. In the scheme 1 $\varepsilon_{\text{corr}}$ slowly increases with N (even or odd) for the ground states regardless of spin value. For example, in the singlet states (for even N) and double states (for odd N) we have

Table 3.

N	2	4	6	8	10
$\varepsilon_{\text{corr}}$	0.170	0.177	0.181	0.183	0.185

Table 4.

N	3	5	7	9
ε_{corr}	0.174	0.184	0.191	0.195

In the parametrization scheme II ε_{corr} becomes less with enlarging N only for the singlet state, albeit the total energy per atom (π-center) increases [9]. For all s the value ε_{corr} tends to the same value if $N \to \infty$.

Somewhat analogous picture is observed for the scheme III which even more than the previous scheme II neglects long-range coulombic interaction of electrons sited at two different centers μ and ν. Besides, such a complete disregard of the coulomb-tail adds up to a distinct behaviour of some characteristics determined by a linear response of the N-electron systems to the external field. The dipole polarizability is a quantity of that type. For numerical results for the ground state of polyenes see Table 5.

Table 5. Average polarizability per one bond (in Å)

	Parametrization scheme		
N	I	II	III $(U = 1.57)$
2	0.89	0.59	0.89
4	0.93	0.60	1.37
6	1.12	0.71	2.15
8	1.31	0.81	3.08
10	1.47	–	–

Clearly, the utter ignorance of the coulombic fail in the Hubbard model predetermines abnormally rapid increase of the polarizability with the length of polyene chain. On this account, when studying dynamical properties (susceptibilities) there seems to be similar discrepancy between real electronic systems and the modelled Hubbard-like ones.

References

1. Luzanov A.V., Physics of Many-Particle Systems **16** 53 (1989)
2. Luzanov A.V., Physics of Molecules **10** 65 (1981)
3. Amos M., Woodward M, J. Chem. Phys. **50** 119 (1969)
4. Knowles P.J., Handy N.C., Chem. Phys. Lett. **111** 315, (1984)
5. Luzanov A.V., Theor. Exp. Chem. **25** 1, (1989)
6. Luzanov A.V., Theor. Exp. Chem. **17** 293, (1981)
7. Nagaoka Y., Phys. Rev. **147** 392, (1966)
8. Luzanov A.V., Pedash Yu.F., Theor. Exp. Chem. **18** 8,(1982)
9. Soos Z.G., Ramasesha S., Phys. Rev. B **29** 5410 (1984)
10. Ramasesha S., Soos Z.G., Synthetic Metals **9** 283, (1984)

Dynamical Correlation in Finite Polymethine Chains

G.G. Dyadyusha and I.V. Repyakh

Institute for Organic Chemistry, 252 130 Kiev, USSR

According to the triad theory of Dähne based on Huckel molecular orbital (HMO) theoretical and experimental results the diversity of unsaturated organic compounds can be interpreted in terms of intermediates between three ideal states: the aromatic state, the polyene state and the polymethine state. All these three states are realized in finite polymethine cyclic or open chains, which contain a methine group CH as elementary link and arbitrary terminal groups (in open chains).

The ideal aromatic state is realized in cyclic hydrocarbons e.g. in [N] annulenes, where $N = 4n+2$. Typical features are identical π–electron bond densities between all neighbor atoms and identical π–electron densities of unit at all atoms.

The ideal polyene state is characterized by alternating double and single bonds in chain–shaped compounds and have similarly aromatics a π–electron density per atom of approximately unit. On light absorption and transition from the ground state to the first excited singlet state a maximum change in the π–bond densities occurs in such manner, that double bonds acquire increased single–bond character and vice versa. Both aromatics and polyenes possess a finite energy gap.

The ideal polymethine state just the aromatic state, is characterized by identical π–electron bond densities. In contrast to aromatic and polyene states, polymethine feature is a strongly alternating π–electron density distribution along the polymethine chain which redistributes maximally at light excitation from the ground state to the first excited singlet state where atoms with a negative excess charge acquire a positive charge and vice versa and polymethines possess zero energy gap when $n \to \infty$, i.e. first π–electron transition wavelength is linear function of chain carbons number. Ideal polymethine state is realized in polymethine dyes (PMD):

$$[X - (CH - CH)_n - CH - Y]^z \, , \quad z = 0, \pm 1 \, , \quad n \leq 20 \, .$$

Physicists often consider aromatics as cyclic polyenes and refer then to one class. In present work we pay attention to polymethine and aromatic common features. We deal with closed–shell compounds. First, lowest π–electron transition wavelength is proportional to π–perimeter system atom number in [N] annulenes with $N < 20$. Secondly, in HMO treatment on factorization of Huckel determinant relative to reflection plane, bisecting opposite bonds one can obtain ideal polymethine secular equations. Lowest π–electron transitions both in cyclic and open polymethine chains (PMC) were considered by means of the Pariser–Parr–Pople (PPP) method with the Hamiltonian:

$$H = \sum_{pq}\sum_{\sigma} H_{pq} a^+_{p\sigma} a_{q\sigma} + \sum_{pq}\sum_{\sigma\eta} \gamma_{pq} a^+_{p\sigma} a^+_{q\eta} a_{q\eta} a_{p\sigma} , \qquad (1)$$

where $a^+_{p\sigma}$ and $a_{p\sigma}$ are the creation and annihilation operators involving σ–spin in the state with the basic atom function φ_p. Accordingly to works [1, 2] it is enough to describe annulene absorption spectra such parametrization of HMO Hamiltonian:

$$H_{pq} = \alpha \delta_{pq} + \beta(\delta_{p,q+1} + \delta_{p,q-1}) ,$$

$$\gamma_{pq} = \gamma_{11}\delta_{pq} + \gamma_{12}(\delta_{p,q+1} + \delta_{p,q-1}) , \qquad (2)$$

here δ_{pq} – Kroneker symbol, α and β – Coulomb and resonance integrals respectively, γ_{11} and γ_{12} – Coulomb electron–electron repulsion integrals on the same and nearest atoms respectively. The parameters of importance are β, γ_{11} and γ_{12}. Besides it is enough to consider only four excited configurations built up with participation of two pairs of degenerate boundary molecular orbitals. Taking a symmetry D_{Nh} for model neutral annulenes in such approximation we obtained lowest annulene transition energies for whole neutral annulene series:

$$E(^1B_{2u}) = -4\beta^0 \sin(\pi/N) + (2/N)\gamma_{12}(1 - \cos(2\pi/N)) ,$$
$$E(^1B_{1u}) = -4\beta^0 \sin(\pi/N) + (2/N)(\gamma_{11} - \gamma_{12}(3 - \cos(2\pi/N))) , \qquad (3)$$
$$E(^1E_{1u}) = -4\beta^0 \sin(\pi/N) + (1/N)(\gamma_{11} + 2\gamma_{12}(1 + 2\cos(2\pi/N))) .$$

In a such way two doubly degenerated annulene ions transition energies were obtained differing for cyclic chains with even ($N = 4n$) and odd ($N = 4n + 1$, $N = 4n + 3$) atom numbers; also triplet transition energies were obtained. In respect that PMD absorption wavelenght is linear functions of atom number, let us linearize the annulene absorption wavelenght formulas, considering $N \gg 1$:

$$\lambda(^1B_{2u}) = hc(-4\pi\beta^0)^{-1}N ,$$
$$\lambda(^1B_{1u}) = hc(-4\pi\beta^0)^{-1}N ,$$
$$\lambda(^1E_{1u}) = hc(-4\pi\beta^0)^{-1}N , \qquad (4)$$

$$\lambda(^3B_{2u}) = hc(-4\pi\beta^0)^{-1}N \;,$$
$$\lambda(^3B_{1u}) = hc(-4\pi\beta^0 - 2\gamma_{11} + \gamma_{12})^{-1}N \;,$$
$$\lambda(^3E_{1u}) = hc(-4\pi\beta^0 - \gamma_{11} + 2\gamma_{12})^{-1}N \;,$$

for $N = 4n + 2$ and

$$\lambda(^1E_1) = hc(-4\pi\beta^0 + \gamma_{11} + 6\gamma_{12})^{-1}N \;,$$
$$\lambda(^1E_2) = hc(-4\pi\beta^0 + \gamma_{11} - 2\gamma_{12})^{-1}N \;, \qquad (5)$$
$$\lambda(^3E) = hc(-4\pi\beta^0 - \gamma_{11} + 2\gamma_{12})^{-1}N \;,$$

for $N = 4n + 1$, $N = 4n + 3$, $N = 4n$, where h–Planck constant, c–light velocity in vacuum.

Approximating neutral annulene experimental wavelength $^1B_{2u}$ and annulene ions ones $^1E_1$ and $^1E_2$ by straight lines, bisecting the beginning of co–ordinate (see Fig. 1), one find from the formulas (4) such parameter values:

$$\beta^0 = -2.34\,\mathrm{eV}\;, \quad \gamma_{11} = 8.68\,\mathrm{eV}\;, \quad \gamma_{12} = 1.86\,\mathrm{eV}\;. \qquad (6)$$

Taking into account parameters (6), we obtain for the [N] annulene lowest triplet transition energy the value $E(^3B_{1u}) = 20/N$ which is somewhat lower that the value $E(^3B_{1u}) = 24/N$ obtained on the base of exact solution for $N \gg 1$ in [6]. The theoretical benzol triplet transition energy $E(^3B_{1u}) = 3.6\,\mathrm{eV}$ (while experimental value is $3.7\,\mathrm{eV}$).

Accordingly to [7] transitions of symmetry $^1B_{2u}$ and $^3B_{1u}$ are quasi-homeopolar and refer to gapeness branch of excited states spectra. The shift to lower wavelength appear in the 18 annulene spectra for quasiionic transitions of symmetry $^1B_{1u}$ and $^1E_{1u}$. So, depending upon absorption spectra type [N] annulenes divide into two classes – neutral molecules with electron density per atom equal to unit and ions with electron density per atom differ from unit, independently of sign and value of deviation. The fact confirms, that $[4n+2]^{4+}$ annulene–cation unlike neutral compound has absorption spectra described by formulas (4).

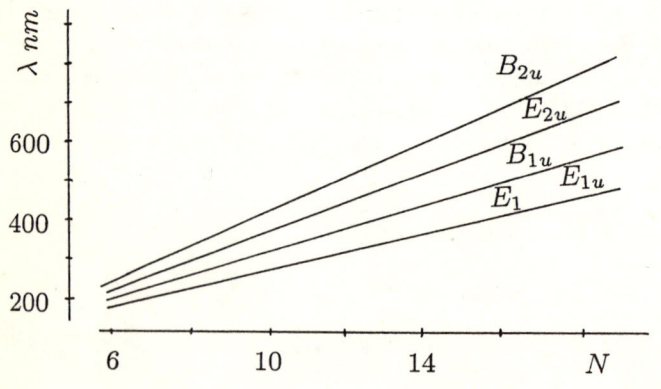

Fig. 1.

Using recently proposed quasi–long dyes approximation [8] to additive productive functions method [9] based on HMO treatment, where $\gamma_{pq} = 0$, the influence of the end–group (EG) topology upon electron structure of the PMD can be studied. In such way it is enough to characterize end–group by means of two parameters. First of them is electron donor ability Φ_0 – one–electron wave phase, which is generated by end–group and determines π–electron density quantity which PMC get from EG. Second one is an effective length L of EG, which determines EG contribution to states density in terms of periodic links (CH–CH) number n. These parameters Φ_0 and L are topological indices and Graph theory enables then to be calculated without solving secular equations. Effective length of the EG was used for analytical wavelength formula to be written:

$$\lambda = (n + L)V , \qquad (7)$$

where V is vinyleneshift, i.e. value which characterized periodic (vinylene) group contribution to states density. Upon this time this value is considered to be the constant equal to 100 nm. But analysis of the experimental data show that deviations in vinyleneshift values can achieve 50%. PPP–calculations for model compounds series where EG was modelled by heteroatoms with different electronegativity that enables Φ_0 variation show that vinyleneshift achieved maximum in PMC with the EG of the middle electron donor ability ($\Phi_0 = 45°$) and decrease when Φ_0 deviate from the middle value. The variation of the electronegativity was obtained by variation of Core ($-U_p$) and Coulomb (γ_{pp}) integrals:

$$U_p = U_p^0 - k\Delta , \quad \gamma_{pp} = \gamma_{pp}^0 + k\Delta/3 , \qquad (8)$$

where $\Delta = 4\,\text{eV}$, $k = 0, 1, 2, 3, 4, 5, 6$, $U_p^0 = -12\,\text{eV}$, $\beta^0 = -2.32\,\text{eV}$. The results of calculations are collected in Table 1.

Table 1. Vinyleneshift in PMD calculated via PPP method depending on vinylene group number, nm

EG	1	2	3	4	5	6
1	61.9	69.3	74.6	77.7	80.4	
2	72.8	81.1	83.7	85.2		
3	90.4	86.3	88.6			
4	95.4	88.1	89.3	88.5	90.2	90.9
5	94.4	87.4	86.8	85.7	86.1	86.4
6	91.5	85.1	83.1	81.9	81.6	81.6
7	89.0	83.1	80.4	79.2	78.8	78.5

The numeration is chosen in such way, that Φ_0 decreases with number increase. EG 4 is of the middle electron donor ability. These results are of qualitative character, as in PPP treatment Φ_0 is not defined. To reveal the feature of the dependence of the periodic link contribution into states density upon EG, it is enough to account the dynamical correlation in open chains as it was done for [N] annulenes, involving the wave functions obtained in quasi–long chain approximation to additive productive functions method. It should be noted that Pople formulas used to determine [N] annulene transition energies are valid only for self–consistent wave functions.

However, from the comparative PMD electronic structure calculations in quasi–long chain approximation, HMO and self–consistent field PPP and CNDO/2 approximations it follows that one–electron wave functions obtained in these approximations are very similar. In chosen approximation singlet and triplet π–electron transition energies are expressed:

$$^1E = \Delta\varepsilon + (\gamma_{11} - 2\gamma_{12})(2 + f)/4L , \qquad (9)$$

$$^3E = \Delta\varepsilon - (\gamma_{11} + 2\gamma_{12})/2L - (\gamma_{11} - 2\gamma_{12})f/4L , \qquad (10)$$

$$f = \sin 2\Phi \cos 2\Delta\Phi/(2\Phi - \pi) , \quad L = L_x + L_y ,$$

$$\Phi = \Phi_{0x} + \Phi_{0y} , \quad \Delta\Phi = \Phi_{0x} - \Phi_{0y} ,$$

here $\Delta\varepsilon$ is the difference between one–electron boundary levels. In chosen model:

$$\Delta\varepsilon = \pi|\beta|/L , \quad \beta = \beta^0 - P_{12}\gamma_{12}/2 ,$$

here P_{12} – carbon–carbon π–bond order.

We should note that expressions (9, 10) are valid only for symmetrical PMD ($X = Y$), while in unsymmetrical ones π–bonds alternation becomes essential. Under this condition we obtain following vinyleneshift formulas:

$$^1V = hc/(\pi\beta + (\gamma_{11} - 2\gamma_{12})(2 + f)/4)^{-1}, \qquad (11)$$

$$^3V = hc/(\pi\beta - (\gamma_{11} + 2\gamma_{12})/2 - (\gamma_{11} - 2\gamma_{12})f/4)^{-1} . \qquad (12)$$

Formulas (11) analysis shows that periodic group contribution to states density achieves maximum in the symmetrical PMC with EG of the middle electronic donor ability ($\Phi_{0x} = \Phi_{0y} = 45°$), i.e. in the case of ideal polymethines; the minimum is achieved in the PMC with the EG of the extremal electronic donor ability ($\Phi_{0x}, \Phi_{0y} = 0°, 90°$). Thus, ideal polymethine transition energy equals to average lowest transition energies of the [N] annulene with double quantity of carbons. The periodic group contribution into the triplet states density is higher that into the singlet one. Substituting parameters (6) into formulas (11) and (12) we find the following extremal values for ideal polymethine singlet and triplet vinyleneshifts:

$$^1V_{\max} = 118\,\text{nm} , \quad ^1V_{\min} = 106\,\text{nm} , \quad ^3V_{\min} = 290\,\text{nm} .$$

These values are consistent with experimentally discovered tendency to decrease of vinyleneshifts in PMC with EG of the extremal electron donor ability. PPP and CNDO/S calculations confirm this tendency.

According to [11] for following compounds:

$$[R - C_6H_4 - (CH - CH)_n - CH - C_6H_4 - R]^+ \cdot BF_4^-, \quad n = 1 \div 4,$$

there are following vinyleneshifts:

$$R = N(CH_3)_2, \quad V = (95 + 5)\,nm, \quad \Phi_0 = 14°,$$

$$R = OCH_3, \quad V = (70 + 5)\,nm, \quad \Phi_0 = 11°,$$

$$R = H, \quad V = (60 + 5)\,nm, \quad \Phi = 0°.$$

Thus, taking into account the dynamical correlation in PMC enables us to describe qualitatively the vinyleneshift dependence upon end groups.

References

1. M.R. Blattman, E. Heilbronner, G. Wagniere: J. Am. Chem. Soc. **90** 4786 (1968)
2. J.F.M. Oth, H. Baumann, J.M. Gilles, G. Schroder: J. Am. Chem. Soc. **94** 3498 (1972)
3. W. Grimmer, E. Heilbronner, G. Hohlneicher, E. Vogel, J.P. Weber: Helv. Chim. Acta **51** 225 (1968)
4. E.A. Lalancette, R.E. Benson: J. Am. Chem. Soc. **85** 2853 (1963)
5. P. Hildenbrand, G. Plinke, J.F.M. Oth, G. Schroder: Chem. Berichte **111** 107 (1978)
6. I.A. Misurkin, A.A. Ovchinnikov: Phys. Sol. **12** 2524 (1970)
7. I.A. Misurkin, G.A. Vinogradov, A.A. Ovchinnikov: Theoret. and Experim. Chem. **10** 587 (1974)
8. G.G. Dyadyusha, A.D. Kachkovskii: J. Inf. Rec. Mater. **13** 95 (1985)
9. G.G. Dyadyusha, M.N. Ushomirskii: Theoret. and Experim. Chem. **22** 127 (1986)
10. G.G. Dyadyusha, I.V. Repyakh, A.D. Kachkovskii: Theoret. and Experim. Chem. **20** 372 (1984)
11. V.H. Grif, V.F. Lavrushin: Ukr. Khim. Zh. **53** 866 (1987)

Electronic Structure and Optical Spectra of Transition Metal Complexes via the Effective Hamiltonian Method

A.V. Soudackov, A.L. Tchougreeff and I.A. Misurkin

Karpov Institute of Physical Chemistry, 103 064 Moscow K-64, USSR

A new semiempirical effective Hamiltonian method capable of taking into account d-electron correlations, the weak covalency of the metal–ligand bonds, and electronic structure of ligand sphere for transition metal complexes is described. The technique is characterized by the use of variation wave function which differs from Hartree–Fock antisymmetrized product of molecular orbitals of the whole complex. The method is applied to the calculation of the $d-d$-spectra of some divalent transition metal complexes. The results for $CoCl_6^{4-}$ and $CoCl_4^{2-}$ are presented.

1. Introduction

Specific properties of transition metal complexes (TMC) are conditioned by their electronic structure, namely, by d-electrons of the transition metal ion. They are responsible for absorption bands in optical spectra of TMC, for magnetic properties etc.

To describe the electronic structure of TMC correctly it is necessary to consider both the correlation effects in the d-shell of the central ion and the effects of the covalency of the metal–ligand bonds. In this note we present a semiempirical effective Hamiltonian method which takes into account the two mentioned effects and allows to calculate electronic spectra of TMC.

2. Theory

Let us define the basis of one–electron functions of TMC. In the valence approximation $4s$-, $4p$- and $3d$-orbitals of the metal atom and valence orbitals of the ligand atoms are the basis ones. The chosen basis of atomic orbitals (AO) may be divided into two parts. The first part contains $3d$-orbitals of transition metal atom (d-subsystem). The second part contains $4s$-, $4p$-orbitals of the transition metal atom and valence orbitals of the ligand atoms (ligand subsystem or L-subsystem). Such a division reflects the main feature of electronic structure of TMC, namely, the presence of an isolated group of strongly correlated d-electrons of the metal atom. Now the total Hamiltonian for TMC in the valence approximation can be written in the following form:

$$H = H_d + H_L + H_c + H_r , \qquad (1)$$

where H_d is the Hamiltonian for d-electrons in the field of the atomic cores of TMC, H_L is the Hamiltonian for electrons in the ligand subsystem, H_c and H_r are respectively the Coulomb and the resonance interaction operators between electrons of the two subsystems.

Chemical bonds between the metal atom and the ligands arise mainly from electron transfer between the ligands and the metal $4s$- and $4p$-orbitals whereas d-electrons are involved in the bond formation to a lesser extent. For the majority of the complexes total charge transfer from the d-shell or into the d-shell usually does not exceeds several per cent of the total number of d-electrons in the relevant valence state of the metal ion in TMC [1].

Based upon these considerations we describe the electronic structure of TMC using wave function with the fixed number of d-electrons. The configurations with other numbers of d-electrons (charge transfer states) will be taken into account with use of the Löwdin partition technique [2].

In a line with [2], we consider the *effective* Hamiltonian $H^{\text{eff}}(E)$, which operates in the subspace spanned by the functions with the fixed number of d-electrons n_d and with $n_L = N - n_d$ electrons in the ligand subsystem. Its eigenvalues coincide with those of the exact Hamiltonian (1).

$$H^{\text{eff}}(E) = PH_0P + H_{RR} , \qquad (2)$$

$$H_0 = H_d + H_L + H_c , \qquad (3)$$

$$H_{RR} = PH_r Q(EQ - QH_0Q)^{-1} QH_r P . \qquad (4)$$

Here P is the projection operator on the subspace of functions with fixed number of d-electrons; $Q = 1 - P$.

It has been noticed [3] that the energy dependence of the effective Hamiltonian is weak. We neglect this dependence and consider the Hamiltonian $H^{\text{eff}}(E_0)$, where E_0 is the ground state energy of the Hamiltonian PH_0P. The effective operator $H^{\text{eff}}(E_0)$ corresponds to the second order of the operator perturbation theory with H_r as a perturbation [4].

We restrict ourselves with the simple case of the complexes where excitations in the ligand subsystem are of very high energy and thus their contribution is negligible. The majority of TMC with the ligands having closed electronic shell satisfies this condition. In this case the ground state of the ligand subsystem can be described by a Slater determinant $\Phi_L(^1A_1)$ with zero total spin. Thus the variation wave function Φ_n may be written in the following form:

$$\Phi_n = \left(\sum_k C_k^n |n_d k\rangle \right) \wedge \Phi_L = \Phi_d^n \wedge \Phi_L \,, \tag{5}$$

where $|n_d k\rangle$ are the spin and symmetry adopted n_d-electron wave functions constructed on the metal d-orbitals; C_k^n are the variation parameters. Note that both the spin multiplicity and the point symmetry of the functions of the type (5) are completely determined by the multiplicity and the symmetry of the functions Φ_d^n.

The solution of the variation problem with effective Hamiltonian $H^{\text{eff}}(E_0)$ gives a pair of interconnected equations for the functions Φ_d^n and Φ_L [5]:

$$\begin{cases} H_d^{\text{eff}} \Phi_d^n = E_d^n \Phi_d^n \,, \\ H_L^{\text{eff}} \Phi_L = E_L \Phi_L \,. \end{cases} \tag{6}$$

The effective Hamiltonians for the subsystems are given by:

$$H_d^{\text{eff}} = H_d + \langle \Phi_L | H_c + H_{RR} | \Phi_L \rangle \,, \tag{7}$$

$$H_L^{\text{eff}} = H_L + \langle \Phi_d^n | H_c + H_{RR} | \Phi_d^n \rangle \,. \tag{8}$$

To find the approximate solution of the equations (6) we use the semiempirical approach. We insert into (8) an initial density matrix ρ of the d-subsystem of the form

$$\rho_{\mu\nu} = \delta_{\mu\nu} n_d / 5 \,, \tag{9}$$

which describes a uniform distribution of electrons in the d-orbitals of the metal atom. This leads to the effective Hamiltonian H_L^{eff} which arises from H_L by averaging of the Coulomb interaction H_c over the density matrix ρ. This averaging results in a renormalization of the one electron parameters of the Hamiltonian H_L. In the CNDO approximation for the ligand subsystem the attraction parameters of the electrons on the metal $4s$- and $4p$-orbitals to the metal core and the effective core charge of the metal atom should be renormalized:

$$U_{ss} \to U_{ss} + n_d g_{sd} \,, \tag{10}$$

$$U_{pp} \to U_{pp} + n_d \overline{g}_{pd} \,, \tag{11}$$

$$Z_M \to Z_M - n_d \,. \tag{12}$$

Here $g_{sd} = (ss|dd) - (sd|ds)/2$; $\overline{g}_{pd}$ is the mean value of the integrals $g_{i\mu}$, where $i = 4p_x, 4p_y, 4p_z$ and $\mu = 3d_{z^2}, 3d_{xz}, 3d_{yz}, 3d_{x^2-y^2}, 3d_{xy}$.

We found, that the contribution of the $\langle\langle H_{RR}\rangle\rangle_d$ term to the orbital energies of the ligand subsystem turned out to be insignificant, and neglected this term.

The calculation of the ligand subsystem in the CNDO approximation with the renormalized parameters gives the one electron density matrix P_{kl}, the energies of the molecular orbitals (MO) ε_i, and MO LCAO coefficients c_{ik}. These quantities are used to construct the effective Hamiltonian H_d^{eff} (7). The operator H_d has the form:

$$H_d = U_{dd} \sum_{\mu\sigma} d^+_{\mu\sigma} d_{\mu\sigma} + \sum_{\mu\nu\sigma} V^{\text{core}}_{\mu\nu} d^+_{\mu\sigma} d_{\nu\sigma}$$
$$+ \frac{1}{2} \sum_{\mu\nu\rho\eta} \sum_{\sigma\tau} (\mu\nu|\rho\eta) d^+_{\mu\sigma} d_{\nu\sigma} d^+_{\rho\tau} d_{\eta\tau} , \quad (13)$$

where $d^+_{\mu\sigma}(d_{\mu\sigma})$ are the creation (annihilation) operators for an electron with the spin projection σ in the μ-th real d-orbital; U_{dd} is the attraction parameter of d-electrons to the metal core; $V^{\text{core}}_{\mu\nu} \equiv \langle\mu|V^{\text{core}}|\nu\rangle$ is the matrix element of the interaction of d-electrons with the ligand atom cores; $(\mu\nu|\rho\eta)$ is the two–electron Coulomb integral.

Averaging the operators H_c and H_{RR} over the ground state wave function Φ_L of the ligand subsystem we derive the effective Hamiltonian H_d^{eff} with renormalized one electron parameters:

$$H_d^{\text{eff}} = \sum_{\mu\sigma} U^{\text{eff}}_\mu d^+_{\mu\sigma} d_{\mu\sigma} + \frac{1}{2} \sum_{\mu\nu\rho\eta} \sum_{\sigma\tau} (\mu\nu|\rho\eta) d^+_{\mu\sigma} d_{\nu\sigma} d^+_{\rho\tau} d_{\eta\tau} ,$$

where the effective core attraction parameters for metal d-electrons $U^{\text{eff}}_{\mu\mu}$ contain the corrections originating both from the Coulomb and the resonance interaction with the ligand subsystem:

$$U^{\text{eff}}_\mu = U_{dd} + W^{\text{ion}}_\mu + W^{\text{cov}}_\mu , \quad (14)$$

where

$$W^{\text{ion}}_\mu = \sum_{i \in s,p} g_{\mu i} P_{ii} + \sum_L (P_{LL} - Z_L) V^L_{\mu\mu}$$

is an ionic term and

$$W^{\text{cov}}_\mu = - \sum_j^{(MO)} \beta^2_{\mu j} \left\{ \frac{(1 - n_j/2)^2}{\Delta E_{\mu j}} - \frac{(n_j/2)^2}{\Delta E_{j\mu}} \right\},$$

is a covalence term. Here P_{ii} is the diagonal matrix element of the one–electron density matrix of the ligand subsystem; $P_{LL} = \sum_{l \in L} P_{ll}$ is the electronic population on the ligand atom L; Z_L is the core charge of the ligand atom L; $V^L_{\mu\mu}$ is the matrix element of the operator of the potential energy of d-electron interacting with electron placed on the ligand atom L; $\beta_{\mu j}$ is the resonance integral between the μ-th d-orbital and the j-th ligand MO; n_j is the occupation number of the j-th ligand MO ($n_j = 0$ or 2); $\Delta E_{\mu j}$ ($\Delta E_{j\mu}$) is the energy which is necessary to transfer an electron from the μ-th d-orbital (from the j-th MO) to the j-th MO (to the μ-th d-orbital).

In the present model the d-level splitting parameters which are used in the crystalline field theory [6] are the differences of the corresponding quantities U^{eff}_μ. In the case of the octahedral complexes the splitting parameter $10Dq \equiv U^{\text{eff}}(e_g) - U^{\text{eff}}(t_{2g})$ takes the form:

$$10Dq = (W^{ion}_{e_g} - W^{ion}_{t_{2g}}) + (W^{cov}_{e_g} - W^{cov}_{t_{2g}}) = 10Dq^{ion} + 10Dq^{cov} \ .$$

This allows one to analyze the relations between the ionic and the covalent contributions to the splitting of d-orbitals in the various transition metal complexes.

Our calculations have shown that the resonance interaction gives the largest contribution to the one–electron d-level splitting. Therefore it seems to be the most important interaction between d-electrons and ligand electrons of TMC.

To calculate the excitation energies in the d-shell (in our approximation these are the energies of the low lying excited states of the whole complex) it is necessary to diagonalize the matrix of the H^{eff}_d constructed in the basis of n_d-electron wave functions.

3. Results and Discussion

To demonstrate a validity of the proposed approach we calculate the electronic structure and the $d - d$-excitation energies of transition metal hexafluoroanions MF_6^{4-}, where M = Mn(II), Fe(II), Co(II), Ni(II), complex ion $Mn(FH)_6^{2+}$, hexachloroanions MCl_6^{4-}, where M = Mn(II), Fe(II), Co(II), and tetrachloroanion $CoCl_4^{2-}$. The semiempirical parameters are taken from literature [7–10] but the interatomic resonance integrals $\beta_{\mu k}$ are parameterized in the same manner as in the MINDO/3 method [11]. The values of the pair resonance parameters β^{M-L} are fitted to approximate the energy of the first transition in the $d - d$-spectra of the complexes considered.

Our calculations correctly predict the point symmetry and the spin multiplicity of the ground state of all the complexes considered. This result seems to be nontrivial because the predictions of the ground state multiplicity of TMC in the framework of the usual SCF methods is a difficult problem to which only a little attention is paid. The proposed description of the correlations in the d-shell allows us to solve this problem. The calculated excitation energies are also in fair agreement with experimental data. The results for octahedral $CoCl_6^{4-}$ and tetrahedral $CoCl_4^{2-}$ presented in Table I are demonstration that the pair resonance parameters β^{M_L} are transferable from one complex to another.

Table 1. $d - d$-Spectra for $CoCl_6^{4-}$ and $CoCl_4^{2-}$ complexes

Complex	Transition	E^{calc}, cm^{-1}	E^{obs}[6], cm^{-1}
$CoCl_6^{4-}$	ground state $^4T_{1g} \to {}^4T_{2g}$	5900	6300
	$\to {}^4 A_{2g}$	12700	13000
	$\to {}^4 T_{1g}(P)$	16700	17000
$CoCl_6^{2-}$	ground state $^4A_2 \to {}^4T_1(F)$	5000	5000
	$\to {}^4 T_1(P)$	15300	15000

4. Conclusion

In this note we believed that the local d-electron correlations as well as the resonance interactions between the d-electrons and ligand electrons play an important part in electronic structure of TMC. We have taken into account these ideas by dividing of all valence electrons of TMC into two parts: d-subsystem and "ligand" subsystem containing ligand electrons and $4s$-, $4p$-electrons of metal atom. The wave function of the whole system has been written as a product of wave functions of these two subsystems. The effective Hamiltonians for the subsystems have been derived by the Löwdin partition technique and the variation principle. A fair agreement with experimental data on $d-d$-spectra of TMC has been obtained.

References

1. Klimenko N.M.: *Quantum chemical calculations of transition metal complexes* (VINITI, Moscow 1978) [in Russian]
2. Lowdin P.-O.: *Perturbation Theory and its Application in Quantum Mechanics* C.H.Wilcox Ed. (Wiley, NY 1966)
3. Hubbard J., Rimmer D.E., Hopgood F.R.A.: Proc. Phys. Soc. **88** 13 (1966)
4. Bogolyubov N.N.: *Selected papers* (Naukova Dumka, Kiev 1970) Vol. 2 [in Russian]
5. McWeeny R., Sutcliffe B.T.: *Methods of Molecular Quantum Mechanics* (Academic Press, London 1969)
6. Lever A.B.P.: *Inorganic Electronic Spectroscopy* (Elsevier, Amsterdam 1986)
7. Pople J.A., Beveridge D.L.: *Approximate Molecular Orbital Theory* (McGraw-Hill, N.Y. 1970)
8. Clack D.W., Hush N.S., Yandle S.R.: J. Chem. Phys. **57** 3503 (1972)
9. Bohm M., Gleiter R.: Theoret. Chim. Acta **59** 127 (1981)
10. Di Sipio L., Tondello E., De Michelis G., Oleari L.: Chem. Phys. Lett. **11** 287 (1971)
11. Bingham R.C., Dewar M.J.S., Low D.H.: J. Am. Chem. Soc. **97** 1285 (1975)

Part III

Multiparticle Effects in Kinetics and Magnetism

Magnetic Properties of the Hubbard Model with Infinite Interactions

V.Ya. Krivnov[1], A.A. Ovchinnikov[1] and V.O. Cheranovskii[2]

[1] Institute of Chemical Physics, Kosygin St. 4, 117 334 Moscow V-334, USSR
[2] Kharkov State University, Institute of Chemistry, Kharkov, USSR

The problem of the ground state multiplicity of the Hubbard model with infinite interaction is considered for the systems consisting of weakly bound segments of length n (an n–chain model). It is shown that the total spin of the ground state grows from 0 to $S_{\max}$ with an increase of electron density ρ. For two–chain model the critical value ρ_c is found, above which the ground state is ferromagnetic. For n–chain model with $n > 2$ a "cascade" of the transitions with alteration of total spin takes place.

The discovery of the high–T_c superconductivity have renewed the interest in the study of the model with strong interaction. The simplest of them is the Hubbard model with in finite repulsion described by the Hamiltonian

$$H = t \sum_{\langle ij \rangle \sigma} (c_{i\sigma}^+ c_{j\sigma} + c_{j\sigma}^+ c_{i\sigma}) + \sum_i n_{i\alpha} n_{i\beta} , \qquad (1)$$

where $U = \infty$. At $U = \infty$ this Hamiltonian is reduced to the first (hopping) term but acting on the space with no double occupied sites. One of the most interesting problem relating to this model is finding the dependence of the ground state multiplicity $S(\rho)$, where S is a total spin of the ground state and ρ is an electron density ($\rho = N/N_0$, where N and N_0 are numbers of electrons and sites).

In a one–dimensional lattice (to be more precise, for the chain with the open ends) the ground state does not depend on the total spin. However, for the finite rings the situation is not trivial, and degeneracy is partially resolved though in a rather specific way [1, 2].

For the two– and three–dimensional model the problem of the ground state multiplicity becomes very complicated and can be analyzed only in two limiting cases $\rho \to 0$ and $\rho \to 1$. In particular, the problem of two electrons can be solved exactly. It turns out that in this case ground state is singlet. On other hand the ground state of Hubbard model on biparticle lattice with a single hole ($N = N_0 - 1$) is ferromagnetic [3] (Nagaoka state). Thus, at intermediate values of ρ there occurs the transition from ferromagnetic state into singlet one. The question arises whether this transition takes place at

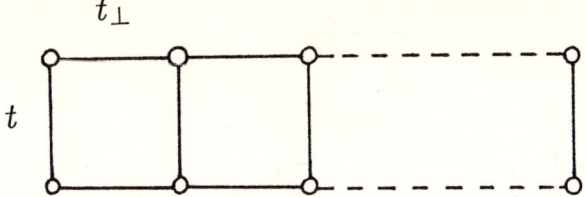

Fig. 1. Two-chain model

finite density ρ or Nagaoka state can be realized for vanishing hole density only. The numerical calculations of the finite clusters [4] show that S is not a smooth function of the electron numbers and do not allow to find the dependence $S(\rho)$ for the macroscopic systems.

We shall study this dependence for quasi–one–dimensional systems of a special type and show that above mentioned transition occurs at finite hole density. At first, let us consider a simple two–chain model which represents a "ladder" (Fig. 1), consisting of two–site segments (dimers) and which is characterized by two transfer integrals t and $t_\perp$. Besides, $\alpha = |t_\perp/t| \ll 1$ which allows use of perturbation theory (PT) in α. For $t_\perp = 0$ energetic levels are spin degenerated. The resolve of degeneracy (up to α^2) leads to an effective spin Hamiltonian H_{eff}, the form of which depends on ρ.

For $\rho \leq 1/2$ ($\rho = N/2L$, L is the number of segments) there can be not more than one electron at each dimer (there is a large (in comparison at with $|t_\perp|$) gap $2|t|$ for the states with two electrons at one segment). It is easy to convince oneself that the action of H_{eff} is reduced to the transfer of electron from occupied to the neighboring empty segment with the transfer integral t and to the exchange interaction of antiferromagnetic type $J(\boldsymbol{S}_i\boldsymbol{S}_{i+1} - 1/4)$ ($J = t_\perp^2/|t|$, $\boldsymbol{S}_i$ and $\boldsymbol{S}_{i+1}$ are the spin operators of the electrons of these segments), if two neighboring segments are occupied. Evidently, these processes are described by the Hamiltonian of one–dimensional $t - J$ model which is equivalent at $J \ll t$ to Hubbard Hamiltonian. For considered system the effective parameters of this Hamiltonian are

$$\rho_{\text{eff}} = 2\rho \; ; \quad t_{\text{eff}} = t_\perp \; ; \quad U_{\text{eff}} = 4t \; . \tag{2}$$

For $\rho < 1/2$ and $\alpha \ll 1$ this equivalence allows to write H_{eff} in the form [5]

$$H_{\text{eff}} = -L|t_\perp|\pi^{-1}\sin\pi\rho + t_\perp^2/|t|\rho(1-(2\pi\rho)^{-1}\sin 2\pi\rho)\sum^{N}(\boldsymbol{S}_i\boldsymbol{S}_{i+1} - 1/4) \; . \tag{3}$$

The ground state of (3) is a singlet. Thus a ladder has the singlet ground state at $\rho \leq 1/2$.

When $1/2 < \rho < 1$ there can be one or two electrons (a "one" or a "pair") at each segment. The wave function of the system containing M

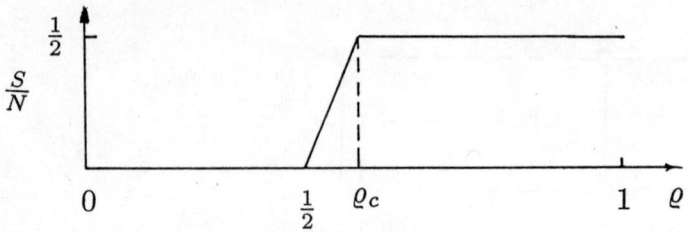

Fig. 2. The dependence $S(\rho)$ for ladder

pairs and $(L-M)$ ones depends on $(L-M)$ spin variables $\sigma_i = 1/2$ and M variables $\lambda_i = 1, -1, 0, \bar{0}$, corresponding to values S and S_z of pairs (1,1;1,-1;1,0;0,0). Hamiltonian H_{eff} is determined by its action upon spin variables of neighboring segments:

$$H_{\text{eff}}\Psi(\ldots\lambda_i,\sigma_j\ldots) = -|t_\perp|(\mathbf{L}_i\mathbf{S}_j + 1/2)\Psi(\ldots\sigma_j,\lambda_i\ldots), \quad (4)$$

where $\mathbf{L}_i$ is the spin 1 operator

$$H_{\text{eff}}\Psi(\ldots\sigma_i,\sigma_{i+1}\ldots) = t_\perp^2/|t|(\mathbf{S}_i\mathbf{S}_{i+1} - 1/4)\Psi(\ldots\sigma_i,\sigma_{i+1}\ldots), \quad (5)$$

$$H_{\text{eff}}\Psi(\lambda_i,\lambda_{i+1}) = 0. \quad (6)$$

Energy (4) is minimal at maximal total spin ($S = 3/2$) of two neighboring segments, while (5) results in the singlet state. Thus (4, 5) describe competing interactions of ferro- and antiferromagnetic types.

At $M = L - 1$ (one hole), the ground state is ferromagnetic [1]. With $M = 1$ (one pair) the competition of (4) and (5) leads to the formation of the region with ferromagnetic ordering (a polaron), i.e. in the region of size L^* round pair, the latter orientates the spins of ones parallel to its own spin. Minimizing the total energy of the system in L^*, we obtain the following expression

$$L^* = (2\pi^2/C)^{1/3}, \quad C = \alpha \ln 2. \quad (7)$$

The variational calculations show that when there are M pair, they gather into one polaron. It is occurs [1], that the size of this polaron linearly increases with M:

$$L^* = \pi M(3\pi C/2)^{-1/3}. \quad (8)$$

When

$$M = (L/\pi)(3\pi C/2)^{1/3}$$

the polaron length is equal to the ladder length L and the total spin $S = S_{\max} = N/2$.

Thus, at ρ equal to

$$\rho_c = 1/2 + (2\pi)^{-1}(3\pi C/2)^{1/3} \quad (9)$$

the system becomes saturated ferromagnet. The dependence of the ground state total spin $S(\rho)$ is shown in Fig. 2.

Calculating the function $S(\rho)$ we neglected the polaron movement and the existence of the boundary region. To estimate the influence of these effects, we have carried out numerical calculations of ladders with $L = 4$ to 11, containing a single pair and estimated critical value α^* under which for a fixed L the decrease of the system total spin from S_{max} to $S_{max} - 1$ takes place. These data [1] are well approximated by the formula

$$L + 1 = 3.07(\alpha^*)^{-1/3} . \qquad (10)$$

The comparison of (10) with (9) testifies to the fact that effects connected with polaron movement are small, and to a good approximation it can be considered as a stationary object with sharp boundaries.

Besides we calculated the ground state energy of ladder with $L = 7$ and $t_\perp = 0.1t$ for different values of total spin S. Our data show that the ground state total spin $S = 0$ (1/2) for even (odd) $N \leq 7$; $S = 2$ for $N = 8$ and $S = N/2$ for $N = 9 - 13$. This confirms the polaron picture as well and coincide with dependence in Fig. 2.

Let us consider now more complicated systems with the topology of the strips formed by interacting segments of length n (an n – chain model). The effective spin Hamiltonian of two neighboring segments, containing one electron each, has the form

$$H = J[(12) - 1] , \qquad (11)$$

where $(12) = 2\boldsymbol{S_1 S_2} + 1/2$, $J \sim n^2 t_\perp^2 / |t|$. (The PT applicability condition limits the length of segments by condition $J \ll |t|$). Hamiltonian (11) has the form (5) and its ground state is singlet. The effective spin Hamiltonian of the interactions "one"–"pair", "pair"–"pair" and etc. for $n > 2$ cannot be simply written in terms of total spin operators of the segments (like pairs of dimer model). The most convenient form of this Hamiltonian can be obtained when one uses cyclic permutations of one–electron spin variables σ_i [6]. For example, the Hamiltonian of the interaction "one"–"pair" has the form:

$$H_{ij} = J_1[(12) + (23)] + J_2[1 + (12)(23)] , \qquad (12)$$

where i, j are numbers of neighboring segments containing one and pair respectively, J_1 and J_2 are effective exchange integrals (for $n = 3$ $J_1 = -(9/16)|t_\perp|$, $J_2 = (1/16)|t_\perp|$).

The calculation shows that the ground state spin of (12) is maximal ($S = 3/2$) and its energy is equal to $-|t|$ for arbitrary value of n.

The Hamiltonian of the interaction of pairs has more complicated form [2], and we cannot find the ground state spin of interacting pairs for arbitrary values of n. The numerical calculations of two–segments strips for $n = 3 - 6$ (Table 1) show, that the ground state of two pairs is singlet in all cases.

Table 1. The ground state energy of two interacting n–site segments (in units $t_\perp^2/|t|$)

$N\backslash n$	3	4	5	6
4	$-0.994(S=0)$	$-1.845(S=0)$	$-2.554(S=0)$	$-3.406(S=0)$
6	0	$-1.401(S=0)$	$-1.857(S=0)$	$-2.675(S=0)$
8	–	0	$-1.766(S=0)$	$-1.943(S=0)$

Now let us go back to finding the dependence $S(\rho)$ the strips with arbitrary values of n. For $\rho \leq 1/n$ there can be not more than one electron at each segment and H_{eff} is reduced to the Hamiltonian of one dimensional $t-J$ model of type (3) with singlet ground state. For $1/n < \rho < 2/n$ there is competition of two types of interactions (ferro– and antiferromagnetic types) as in dimer model, leading to the polaron picture. It is necessary to have in mind only that at $\rho \gtrsim 1/n$ the polaron is formed by pairs amongst the background ones and at $\rho \lesssim 2/n$ on the contrary by ones amongst the background pairs. At $\rho = 2/n$ ground state is singlet. Such the strip is saturated ferromagnet when

$$1/n - (3\pi\varepsilon_1/2|t_\perp|)^{1/3}(n\pi)^{-1} < \rho < 2/n - (3\pi\varepsilon_2/2|t_\perp|)^{1/3}(n\pi)^{-1} \,, \quad (13)$$

where ε_1 and ε_2 are energies (on segment) of the ground state of phases, consisting of the ones or pairs only. As an example, for

$$n=3\,, \qquad \varepsilon_1 = 7(2)^{1/2}\ln 2 t_\perp^2/3|t|\,, \qquad \varepsilon_2 = 0.83 t_\perp^2/|t|\,.$$

The behaviour of $S(\rho)$ at $\rho > 2n$ is associated with the competition of the interactions "pair"– "three", "three"– "three" and etc. (the "three" is 3 electron complex), if such competition takes place.

To reveal this we calculated the spectra of two–segment strips with $N = 5, 7$ and $n = 3-6$ (Table 2).

Table 2. The ground state energy (in units $|t_\perp|$) for different value of total spin S of two interacting n–site segments containing 5 (a) and 7 (b) electrons

(a)

$n\backslash S$	5/2	3/2	1/2
3	-1	-0.8854	-0.7863
4	-1	-0.8784	-0.8164
5	-1	-0.8927	-0.8425
6	-1	-0.8978	-0.8525

(b)

$n\backslash S$	7/2	5/2	3/2	1/2
4	-1	-0.9488	-0.8970	-0.8245
5	-1	-0.9377	-0.8889	-0.8324
6	-1	-0.9415	-0.8989	-0.8417

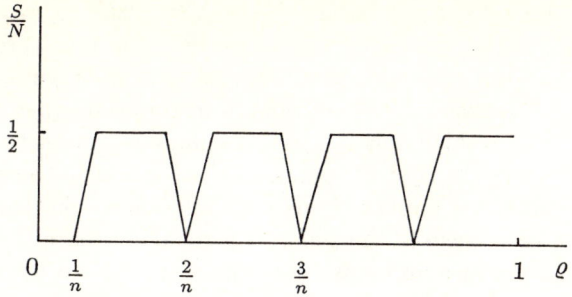

Fig. 3. The dependence $S(\rho)$ for n–sites model

In all cases the ground state spin is maximal. Analogous calculations for even N (Table 1) show that the ground state of two–segments with $N = 4-8$ and $n = 3 - 6$ is singlet except the case $N = 8$ and $n = 5$, where $S = 3$. These results allow to assume that the ground state total spin of k and $k-1$ electrons is maximal, whereas the interaction of two segments with equal number of electrons leads to the state with $S < S_{\max}$. As in dimer model we can say about the polarons formed by threes amongst the background pairs and etc, that leads to the dependence $S(\rho)$ which is shown schematically in Fig. 3.

Thus, a "cascade" of the transitions with alteration of total spin takes place for the strips consisting of weakly bound segments with $n > 2$.

Up to this point we considered in essence one–dimensional systems. Equally to them more complicated systems consisting of dimers packed up two- or three–dimensional lattice can be considered. The construction of H_{eff} for such systems can be done in a manner analogous to the way this was considered above. For $\rho < 1/2$ H_{eff} is equivalent to Hamiltonian of the $t - J$ model with parameters from (2). The magnetic properties of these models have been discussed by Visscher [7] and Larkin and Ioffe [8]. According to these works the phase diagram of ground state for $\rho < 1/2$ has form showed in Fig. 4.

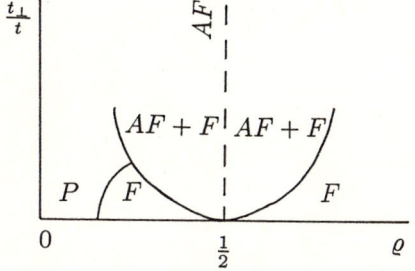

Fig. 4. The ground state phase diagram of two– and three–dimensional dimer models

Ferromagnetic phase is Nagaoka state. In the region AF+F the separation into ferromagnetic and antiferromagnetic phases occurs (the holes are in ferromagnetic phase and do not penetrate into remaining antiferromagnetic part of the system). At decreasing of ρ the transition to paramagnetic phase occurs. However, up to now this transition is not examined quantitatively. At $\rho = 1/2$ considered system is antiferromagnet. At $\rho > 1/2$ there are pairs and ones and the polarons appear though in difference from the case $\rho < 1/2$ the ferromagnetic phase is the polaron formed by pairs. At last, at further increasing of ρ the system transfers to the state of saturated ferromagnet.

In conclusion we note, that for the considered systems the transition from ferromagnetic to the singlet ground state occurs at finite value of ρ though character of this transition is different for different systems.

References

1. V.Ya. Krivnov, A.A. Ovchinnicov, V.O. Cheranovskii: Synthetic Metals **33** 65 (1989)
2. V.Ya. Krivnov, A.A. Ovchinnikov, V.O. Cheranovskii: Teor. Mat. Fiz. **82** 216 (1990)
3. Y. Nagaoka: Phys. Rev. **147** 392 (1966)
4. M. Takahashi: J. Phys. Soc. Japan **51** 3475 (1982)
5. D.J. Klein, W.A. Seitz: Phys. Rev. B **10** 3217 (1974)
6. V.Ya. Krivnov, A.A. Ovchinnicov, V.O. Cheranovskii: (this issue)
7. P.B. Visscher: Phys. Rev. B **10** 943 (1974)
8. L.B. Ioffe, A.I. Larkin: Phys. Rev. B **37** 5730 (1988)

Anomalous Transport Through Thin Disordered Layers

S.F. Burlatsky[1], *G.S. Oshanin*[1] *and A.I. Chernoutsan*[2]

[1] Institute of Chemical Physics, Kosygin St. 4, 117334 Moscow V-334, USSR,
[2] Department of Physics, Gubkin's Gas and Oil Institute, Leninsky pr. 65, 117296 Moscow, USSR

A wide variety of superconductive and semiconductive technique problem lead to the investigation of conductivity of disordered thin layers. This problem may be reduced via Einstein relation to the description of particles diffusion in corresponding inhomogeneous media. Since the disordered thin layers are the mesoscopic objects it is natural to expect the correlation effects to be of crucial importance in such systems. The results of some recent experimental studies [1–2] indicate that the transfer processes in spatially inhomogeneous thin layers are supported by the rare random channel (which occur due to the fluctuations in the spatial distribution of inhomogeneities) with anomalously high conductivity. In this paper we present the theoretical investigation of such fluctuation effects in disordered thin layers within the framework of diffusion through thin inhomogeneous membrane [3–4].

We consider the model of three–dimensional barrier membrane containing randomly distributed repulsive impurities (Fig. 1) as a function of the membrane's thickness L.

The membrane's diffusive permeability is defined as

$$\chi = \frac{L}{D\Delta c_{\rm ph}} \langle J \rangle , \qquad (1)$$

where D is the diffusion coefficient in a membrane without impurities, $\Delta c_{\rm ph}$ is the difference between substance concentrations in phases outside of the membrane on both sides of it, and $\langle J \rangle$ is the mean diffusive flux (averaged over different points on the membrane's surface or over different realizations of a membrane with impurities). The steady flux through the homogeneous membrane with $U(\mathbf{r}) = 0$ is equal $J_0 = D\Delta c_{\rm ph}/L$ (Fick's law, χ is independent on L); we predict the anomalous (more strong that $1/L$) dependence of χ upon L in a case of the disordered membrane.

We demonstrate that fluctuations in the distribution of impurities enhance the value of $\chi(L)$ and cause the anomalous behavior as compared to the Fick's law. We show that for thin membranes ($L < L^*$) the permeability χ drops as $\exp(-L)$; for intermediate thicknesses ($L^* < L < L^{**}$) the

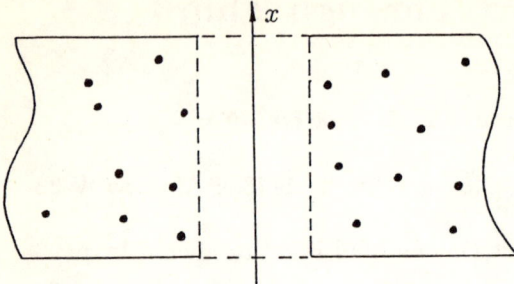

Fig. 1. A three–dimensional membrane containing randomly distributed repulsive impurities. The dashed lines parallel to the X–axis represent a random impurity- free channel that spans the membrane

stretched exponential dependence is valid, $\chi \sim \exp(-L^\gamma)$, $\gamma < 1$; and for $L > L^{**}$ the permeability χ approaches constant value, i.e. the Fick's law is valid. The borderline thicknesses L^* and L^{**} are functions of temperature ($L^* \to \infty$ and $L^{**}/L^* \to \infty$ as $T \to 0$), impurity concentration and parameters of repulsive potentials.

Let us consider a $3d$ membrane containing randomly distributed immobile repulsive centers (Fig. 1). The steady concentration $c(\mathbf{r})$ of particles diffusing through the membrane with a potential $U(\mathbf{r})$ submits to the steady–state Smoluchovsky equation:

$$\mathrm{div}\mathbf{J} = 0 , \quad \mathbf{J}(c(\mathbf{r})) = D(\mathrm{grad} + \beta\mathrm{grad}U(\mathbf{r}))c(\mathbf{r}) , \tag{2}$$

where $\mathbf{J}$ is the steady diffusive flux, D is the diffusion coefficient in a membrane without impurities (homogeneous membrane) and $\beta = 1/kT$. We assume that on both sides of the membrane there are phases with the diffusion coefficient of particles D_{ph} large compared to the diffusion coefficient in membrane, i.e. the local concentrations in the phases are spatially uniform. The local concentration on membrane surfaces is connected with concentrations in the phases via the Boltsman's factors:

$$\begin{aligned} c(0, y, z) &= \exp(-\beta U(0, y, z))c_{\mathrm{ph}} \quad (x < 0) \\ c(L, y, z) &= \exp(-\beta U(L, y, z))c_{\mathrm{ph}} \quad (x > L) , \end{aligned} \tag{2a}$$

which are the boundary conditions to (2)

The random potential is assumed to be equal to

$$U(\mathbf{r}) = \sum_j U_0(\mathbf{r} - \mathbf{R}_j) \quad (0 < x < L, \quad -\infty < y, z < \infty)$$

(and zero outside the membrane). Here $\mathbf{r}$ is the radius–vector of the diffusive particle, $\mathbf{R}_j$ is the radius–vector of the j-th repulsive center. We assume a random distribution of impurities, i.e. the set $\{R_j\}$ is Poissonian. The repulsive potential U_0 is taken either in the multipolar type form

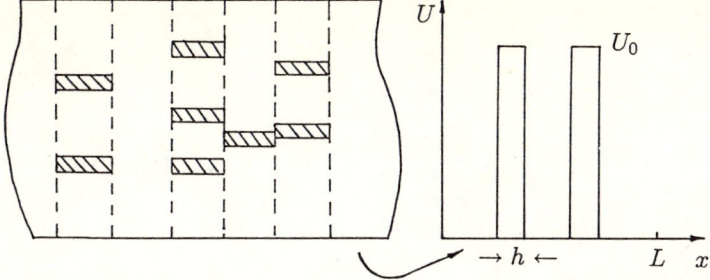

Fig. 2. A model of 3d barrier membrane containing randomly placed high step–like barriers. The dashed lines represent the walls impermeable for the diffusive particles

$$U_0 = u_0 b^m / |r - R_j|^m, \quad m > 3, \tag{3a}$$

or in the exponential form

$$U_0 = u_0 \exp(-|r - R_j|/b), \tag{3b}$$

which mimics hard–core interaction; these are the basic models widely accepted in investigation of the most general properties of disordered media [5].

Let us begin with an illustrative example a simple step–like potential barrier, $U(r) = U$ within the membrane volume and zero in the phase. One can easily show that the (2) yields the Fick's type result

$$\chi = \exp(-\beta U), \tag{4}$$

i.e. χ is independent on L. The straight mean–field approximation consists in using (4) with some effective potential $U = U_{\text{eff}}$. However, this approach is rather rough since we neglect the inhomogeneity in space distribution of repulsive centers. Naturally, at least for thin membranes, the bulk of diffusive flow goes through the channels with anomalously low concentration of repulsive centers (Fig. 1), i.e. with low potential barrier. Qualitatively, it is quite similar to the quantum particles tunneling through disordered layers, when the maximum flux is supported by random channels with unit transparence [5, 6].

In order to illustrate qualitatively this basic physical conception let us consider the following simple quasi–1D model of a membrane with impurities. Consider the membrane of thickness L divided by absolutely impermeable walls into identical straight pores (Fig. 2). These pores contain randomly distributed (with mean linear concentration n) impurities in form of identical step–like narrow barriers with thickness $h(nh \ll 1)$ and high potential level $U_0(\beta U_0 \gg 1)$. From (2) one finds the permeability of the pore with $N(N > 0)$ barriers to be equal $\chi(N) = \exp(-\beta U_0)L/Nh$.

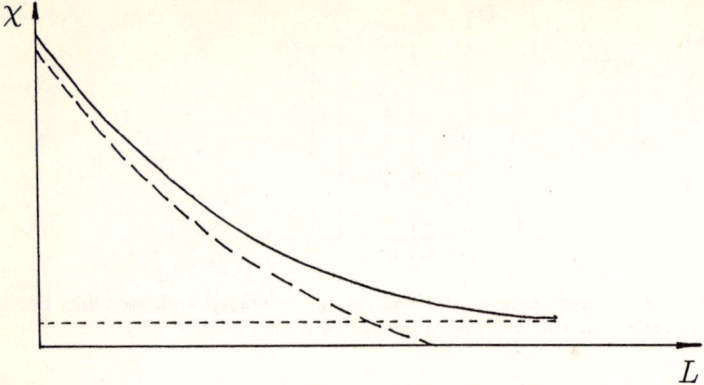

Fig. 3. The dependence of $\langle \chi \rangle$ upon the membrane thickness L. The dashed line corresponds to the exponential dependence caused by the flow through barrier–free channels. The dotted line corresponds to the "mean–field" level. The solid line represents the summery dependence of χ on L

As the permeability of an empty pore is equal 1, at small L the empty pores give the main contribution to $\langle \chi \rangle$; however, since the probability for the existence of empty pores $\exp(-nL)$ drops quickly with L, at larger L the mean permeability should approach the constant mean–field value. Let us write down the exact expression for $\langle \chi \rangle$:

$$\langle \chi \rangle = \exp(-nL) + \exp(-\beta U_0)\frac{L}{h}\langle \frac{1}{N} \rangle , \qquad (5a)$$

where averaging is performed over non–empty pores. The averaging yields $\langle 1/N \rangle = (Ei(nL) - \ln(nL) - C)\exp(-nL)$, where $Ei(x)$ is the integral exponent and C is the Euler constant. For $nL \gg 1$ $Ei(nL) \sim \exp(nL)/nL$, and we obtain

$$\langle \chi \rangle \sim \exp(-nL) + \exp(-\beta U_0/nh . \qquad (5b)$$

For $L \ll L^* = \beta U_0/n$ the bulk flow trough the empty pores and $\langle \chi \rangle$ depends exponentially on L, and for $L \gg L^*$ it approaches the small constant value, i.e. the Fick's law is restored (Fig. 3).

Let us now turn to the $3D$ membrane with impurities. In present paper we restrict our consideration by the qualitative analysis of this problem, using the "optimal fluctuation" type approach [5] (the rigorous results will be presented elsewhere [4]). We begin by defining two characteristic parameters – the mean distance between repulsive centers R_m from the relation $nR_m^3 = 1$ (n is the mean spatial concentration of impurities) and interaction radius R_{int} from the equation $\beta U_0(R_{\text{int}}) = 1$. For multipolar (3a) and exponential (3b) interactions respectively

$$R_{\text{int}} = b(\beta u_0)^{1/m} , \qquad (6a)$$

$$R_{\text{int}} = b\ln(\beta u_0) \ . \tag{6b}$$

We assume that $R_m \ll R_{\text{int}}$, i.e. $nR_{\text{int}}^3 \gg 1$.

We restrict the spectrum of spatial fluctuations by the cylindrical fluctuations of minimal length (Fig. 1), i.e. we consider the impurity–free channels that span the membrane with axis orthogonal to the membrane surfaces. Assuming the homogeneous distribution of repulsive centers outside this cylinder (impurity–free channel), we estimate the mean potential on it is axis:

$$U_{\text{fl}}(R,L) = \langle \sum_j U_0(\mathbf{r} - \mathbf{R}_j) \rangle = n \int_0^L dx \int_R^\infty \rho d\rho U(x,\rho) \ ,$$

where R is the channel's radius. Since the probability for the existence of impurity–free channel of volume V is $\exp(-nV)$, we can estimate the contribution of such fluctuations to the diffusive permeability to be equal $\exp(-n\pi R^2 L - \beta U_{\text{fl}}(R,L))$. Integrating over R by "saddle point" method, we estimate the radius of "optimal fluctuation" R_L (it minimizes $n\pi R^2 L + \beta U_{\text{fl}}$ and the diffusion permeability $\chi(L)$.

If $L \ll R_{\text{int}}$, integration over x is reduced simply to multiplying on L and we get for both repulsive potentials

$$R_L \sim R_{\text{int}} \ , \tag{7a}$$

$$\ln \chi \sim -nR_{\text{int}}^2 L \ , \tag{7b}$$

($L \ll R_{\text{int}} = L^*$), where R_{int} is determined by (6). For L large compared to R_{int} the integration over x in U_{fl} may be expanded to infinity, and U_{fl} does not depend on L. As a consequence, the diffusion permeability follows a stretched–exponential law for multipolar repulsion (3a):

$$\ln \chi \sim -nR_L^2 L \sim -nR_{\text{int}}^3 (L/R_{\text{int}})^{(m-3)/(m-1)} \tag{8a}$$

and an exponential–logarithmic law for exponential repulsion (3b):

$$\ln \chi \sim -nR_L^2 L \sim -nLb^2 \ln^2(\beta u_0 \frac{R_{\text{int}}}{L}) \ . \tag{8b}$$

Let us emphasize that the "optimal fluctuation" radius R_L diminishes with L in this limit ($L \gg R_{\text{int}}$):

$$R_L \sim R_{\text{int}}(R_{\text{int}}/L)^{1/(m-1)} \ll R_{\text{int}} \ ,$$

$$R_L \sim b\ln(\beta u_0 \frac{R_{\text{int}}}{L}) \ll R_{\text{int}}$$

for potentials (3a,b) respectively.

We have estimated the the contribution to $\chi(L)$ of the fluctuation impurity–free channels; this contribution χ_{fl} is represented by (7, 8). Similar to (5), one can expect $\chi(L)$ to be a sum of the fluctuation term χ_{fl} and "mean–field" term χ_{mf}:

$$\chi(L) \sim \chi_{\text{fl}} + \chi_{\text{mf}} .$$

The similar feature (i.e. the additivity of mean–field and fluctuation terms) we observed in the trapping problem [7] and quasi–1D trapping–barrier problem [8]. The "mean–field" permeability is given by (4) with some effective step–like barrier $U = U_{\text{eff}}$. The value of U_{eff} may be estimated as follows. For a given configuration $\{R_j\}$ and corresponding potential surface $U(r)$ we choose the minimal level $U_{\min}$ when the percolation occurs (i.e. for $U > U_{\min}$ the points $U(r) < U$ form the space region that spans membrane). In other words, for $U > U_{\min}$ there exist many "transport paths" along which the potential is less than U. Naturally, these paths may be tortuous and ramified. The weight of transport paths on the potential level U increases as a power of $(U - U_{\min})$, but their contribution to χ drops as $\exp(-\beta U)$. One can expect that transport paths with $U \simeq U_{\min}$ give the bulk contribution to the permeability, i.e. $\langle U_{\min} \rangle$ is of the same order with U_{eff}. On the other band, this percolation level $U_{\min}$ is of the same order with $U(R_m)$ [9, 10], where R_m is the mean distance between repulsive centers. Therefore, the following estimation on the mean–field permeability χ_{mf} is valid:

$$\chi_{\text{mf}} \sim (-\beta U_0(n^{-1/3})) , \qquad (9)$$

where $U_0(r)$ is defined by (3), i.e. one can observe nontrivial dependence of χ_{mf} upon the concentration of impurities.

Evidently, for large L the contribution of impurity–free channels χ_{fl} becomes small (FIg. 4) as compared to the "mean–field" value χ_{mf}, and the expected value of χ must approach the the constant value (9). Let us estimate the borderline value of membrane thickness L^{**}, when the χ_{fl} and χ_{mf} are equal to each other. For multipolar interaction (3a) one gets

$$L^{**} = R_{\text{int}}(R_{\text{int}}/R_m)^{m-1} .$$

and for exponential interaction–

$$L^{**} \sim R_{\text{int}} \frac{U_0(R_m)}{nR_{\text{int}}^3} ,$$

where $U_0(r)$ is determined by (3b).

Therefore, we have the following picture (Fig. 4). There are three different intervals of membrane thickness L with different dependence $\chi(L)$: the exponential dependence (7) for $L \ll L^*$; the more slow stretched exponential dependence (8) for $L^* \ll L \ll L^{**}$; the small constant mean–field value (9) for $L \gg L^{**}$.

Let us describe how the position of these intervals changes with temperature. In the limit of high temperatures $T \gg T_1(nR_{\text{int}}^3(T_1) = 1)\chi$ does not depend on L in the two regions (7, 8), i.e. $\chi(L)$ has a constant mean–field value at all L. In the opposite limit $T \ll T_1$ there are two possibilities. For potential (3a) (all b) and potential (3b) ($b \ll R_m$) for the entire temperature

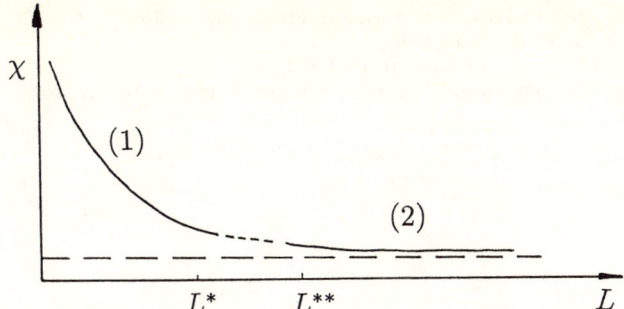

Fig. 4. The dependence of χ upon the membrane thickness L. (1) – the exponential dependence (7), (2) – the stretched exponential dependence (8). The dashed line corresponds to the mean–field level of (9)

interval $T \ll T_1$ there exist both regions (7, 8) (i.e. $L^* \ll L^{**}$), and $\chi(L)$ varies essentially in these regions. In the special case $b \gg R_m$ in potential (3b) there is some temperature interval $T_2 \ll T \ll T_1$, where $L^* \gg L^{**}$, i.e. the exponential dependence (7) changes at $L = L^{**}$ to mean–field dependence (9) without the intermediate interval (8). At $T \ll T_2$ we again have both intervals (7, 8) ($L^* \ll L^{**}$). At $T \to 0$ the intervals (7, 8) of anomalous behavior become infinitely wide ($L^* \to \infty$, $L^{**}/L^* \to \infty$).

To summarize, we have studied the diffusive flow through the inhomogeneous membrane containing repulsive impurities. We have shown that fluctuations in impurity distribution lead to the anomalous exponential dependence of the permeability upon the membrane thickness in the limit of small L. Such behavior is governed by the random impurity-free channels that span the membrane. For larger L the permeability χ dependence on L changes from exponential to stretched exponential law. Finally, at large L limit the permeability χ approaches a constant value corresponding to the diffusive paths on the percolation energy level.

We thank Professors R.J. Rubin, J.B. Hubbard, S. Redner, A.A. Ovchinnikov and S.F. Timashev for helpful discussions.

References

1. V.M. Archangelsky et al.: JETP Lett. **51** 61 (1989)
2. T.D. Gierke et al.: J. Polymer Sci. **19** 1687 (1981).
3. S.F. Burlatsky, G.S. Oshanin, S.F. Timashev: (1990) to be published
4. S.F. Burlatsky, G.S. Oshanin, A.I. Chernoutsan: Phys. Lett. **A** (1990), to be published
5. I.M. Lifshitz, S.A. Gradeskul and L.A. Pastur: *Introduction to the Theory of Disordered Systems* (Wiley, New York 1988)
6. R. Rubin: J. Math. Phys, **11** 1857 (1970)

7. S.F. Burlatsky and A.A. Ovchinnikov: Sov. Phys. JETP **65** 908 (1987)
8. A.I. Onipko: ITF AN UkrSSR Preprint (1986)
9. B.I. Shklovsky and A.L. Efros: Sov Phys. Usp. **18** (1976) 845
10. D. Stauffer: *Introduction to Percolation Theory* (Taylor and Francis, London 1985)

Correlation Effects in Many–Body Reactive Systems

S.F. Burlatsky, G.S. Oshanin and A.A. Ovchinnikov

Institute of Chemical Physics, Kosygin St. 4, 117 334 Moscow V-334, USSR

Charge carrier dynamics in disordered semiconductors has been the subject of many experimental and theoretical studies during the past two decades. Different attempts have been made to describe this phenomenon. Scher and Montroll [1] used the theory of continuous-time random walks to describe carrier motion in amorphous materials; here one assumes a hopping motion between localized sites of a regular lattice, motion governed by a waiting-time distribution. Schmidlin [2] suggested a multiple trapping model, in which the carrier transport is dominated by capture and release processes involving temporal traps. These basic models were often utilized to account distinct degrees of disorder: different intersite distances, energy levels, barrier heights, random transition rates or combinations of all these [3–6]. However, the most of available studies were concentrated on the anomalous transport properties of a single charge carrier.

In this paper we focus on some other aspects of charge carrier dynamics in condensed media. We review some recent results concerning the essentially many–particle correlation–induced dynamics of systems with carriers recombination, trapping and external generation within the framework of the diffusion–controlled processes (DCP) theory. For simplicity we consider only the diffusive motion of carriers. However, all the results presented can be easily generalized on the cases of non–diffusive carrier motion. In general, DCP theory describes the kinetic behavior of many–particle processes, such as, e.g. reactions

$$A + B \longrightarrow C , \qquad (1a)$$

$$A + T \longrightarrow T , \qquad (1b)$$

involving diffusive particles A, B, C and diffusive or immobile traps. Particles react when they approach each other at distance a-reaction radius. The reaction radius a equals the sum of particles radii, $a = R_A + R_B$ (or R_T), for contact reactions and is larger than this sum for distant reactions. The direct reaction – the formation of the reaction product C or capture of A for (1b) occurs with some finite probability determining the direct rate constant K. The backward reaction–unimolecular break–up of C (release from the trap) is defined by the intrinsic constant K_-. This paper is organized as follows: in the first section we briefly introduce some results based on Smoluchowsky mean–field approach. In the second section we present the results concerning the influence of spatial correlations on the long–time reaction kinetics in some simple systems. Finally, in the third section we discuss the reaction kinetics in systems where the spatial correlations are decisive from the earliest times and govern the conversion of the bulk of reactive species.

1. The field of DCP was first stimulated by the work of Smoluchowsky [7] concerning the kinetics of irreversible coagulation of colloid particles. His approach, based on the one–particle diffusion equation with adsorbing boundary leads to the following mass equation for the mean density of particles

$$\dot{C}_A(t) = \dot{C}_B(t) = -K_{\text{eff}} \cdot C_A(t) \cdot C_B(t) , \qquad (2a)$$

where the overdot denotes time derivative and the "effective" rate constant in three–dimensional systems reads (for $4\pi Da \ll K$)

$$K_{\text{eff}} = 4\pi Da \cdot \{1 + a/(\pi Dt)^{1/2}\} , \quad D = D_A + D_B . \qquad (2b)$$

The Smoluchowsky approach is the basis of traditional mean–field methods of DCP investigations [8–11]. In particular, K_{eff} was evaluated for the case $4\pi Da \approx K$ [8]. An analogous to (2b) representation was derived for the effective constant of backward reaction [9]. Much of the emphasis has been put on studying the reaction (1) kinetics in low dimensional systems ($d \leq 2$) and in systems with nonlocal interactions between species. It was shown that in low dimensional systems the effective rate constant becomes rather an effective rate coefficient i.e., K_{eff} becomes a function of time due to the peculiar properties of low dimensional random walks. The potential interactions between reactive species does not affect the long–time dependence of $C(t)$, but lead to the renormalization of the effective rate constants within the framework of the mean–field descriptions [10–12].

2. The Smoluchowsky approach has a major limitation since it ignores the many–particle nature of DCP problems. On the other hand, the course of any reaction leads to the appearance of spatial correlations between the reagents. Within the Smoluchowsky approach only the short–wave two–particle correlations are accounted, but the long–range many–particle correlations caused by reactions between species are neglected and the independence of reaction probability for each reactive pair is assumed. It is worth–while to mention that for a wide variety of many–particle problems of statistical physics the mean–field approximations are quite acceptable. In contrast, the problems where the many–particle behavior is decisive, as a rule, are unsolvable. However, within the last years there seems to have been the considerable progress in the many–particle aspects of DCP theory. It was recently shown that a great majority of DCP are correlation–controlled at large t limit and in the same time a large body of exact results was obtained. In particular, it has been discovered that the trapping reaction (1b) kinetics at $t \to \infty$ is drastically dependent on the spatial distribution of traps T. For

random placement of traps the Smoluchowsky approach, which predicts an exponential decay of A, is invalid and the trapping kinetics exhibits unusual behavior [13-17]

$$\ln C_A(t) \approx -t^{d/d+2} . \tag{3}$$

This stretched exponential decay stems from the presence of large, but very rare, trap-free cavities. In such a cavity the particle life-time is anomalously long and the contribution of these long life-times leads to the decay of eq.(3). It is worth-while to mention that (3) is similar (up to the inverse Laplace transform) to the well-known Lifshitz result concerning the low energy spectrum of electron in array of randomly distributed immobile scatterers [18,13].

The second important instance, where the spatial correlations are decisive, is the strictly bimolecular (SB) irreversible reaction (1a) with $C_A(0) = C_B(0)$. It was shown [15,19-22] that A and B mean concentrations follow an anomalous long-time decay

$$C_A(t) = C_B(t) \approx t^{-d/4} , \tag{4}$$

which displays a slower kinetic behavior for $d < 4$ as compared to the mean-field predictions. The point is that small fluctuation-induced initial A and B concentration difference $z(r,t)$ is not affected by recombination reaction. The diffusive decay of $z(r,t)$ is a slower process than the mean-field recombination decay entailed by (2). For large t the reaction bath becomes separated into the large domains (with characteristic size $R \approx t^{1/2}$ containing particles of only one type and the kinetic behavior is controlled by diffusive "transport" of regions enriched by A species and ones enriched by B to each other, what entails the decay of (4). It means that at $t \to \infty$ limit

$$C_A(t) \approx (-G_{AB})^{1/2} = \langle -(C_A(r,t) - C_A(t))(C_B(r,t) - C_B(t))\rangle^{1/2} ,$$

i.e., the mean-field description (2) is violated and the decrease of densities is governed by pair correlation functions. The correlation-induced decay of (4) was first obtained by means of approximate methods based on the latter physical concept [15,19,20-22]. Its asymptotical validity was rigorously proved in [17,22] and observed in MC simulations [20]. However, these longlive fluctuation states govern the mean density decrease only in case of SB reaction. If the initial densities are not equal to each other the correlation effects are suppressed at large times and the mean-field exponential decay is valid. The mean-field description is also valid for the SB irreversible reaction of charged species [23]. The spatial regions created by density fluctuations induce the local electric fields which faster than diffusion smooth the initial thermal inhomogeneities.

The correlation effects induce the universal long–time behavior of reversible reactions (1). It was shown that a wide variety of reversible bimolecular reactions are defined by power–law approach to equilibrium at $t \to \infty$ [24–29]

$$C_A(t) - C_A(\infty) \approx t^{d/2} , \qquad (5)$$

instead of exponential mean–field decay law. For instance, such kinetic behavior exhibit the reversible binary reaction $A+A\text{---}B$ [24] and more general coagulation/fragmentation reaction [27]. The decay (5) defines the long–time kinetics of strictly bimolecular reversible reaction (1a) of uncharged species [24,25], non–stoichiometric reaction (1a) [28] and reaction (1a) with charged species participation [26,27]. Interestingly, in the latter case the decay (5) is supported by the special type of density fluctuations – the fluctuations of the same sign, i.e. the spatial regions where the deviations from the mean density have the same value and sign both for A and B [27]. These electrically neutral fluctuations are smoothed only by diffusion what causes the power–law approach to the equilibrium.

Summing up the results on the long–time behavior of reversible conversions we deduce [26–28] that the decay (5) is valid for all reversible reactions, since the analog of $z(r,t)$-pure diffusive mode exists for any set of successive and parallel reactions (e.g. the sum of reagents and products local densities).

3. Correlation effects are essentially enhanced in systems prepared by a steady-state external source [29–30]. If the particles A and B are generated randomly and independently of each other with a constant mean rate the irreversible reaction (1a) (and also for the reversible one [26,27]) leads to the appearance of strong correlations in the species distributions. The long–wavelength asymptotic of the fluctuation spectrum changes [26–31] (is not Poissonian) and this leads to the change of the long–time kinetics of (1a) after the source is switched off, $C_A(t) \approx t^{-1/4}$ for $d = 3$ [30]. Interestingly, the approach of $C_A(t)$ and spatial correlators to equilibrium obeys the fluctuation–induced law $C_A(\infty) - C(t) \approx t^{-1/2}$ [29] in 3d, instead of exponential mean–field prediction. Similar to the considered above systems the nontrivial relaxation laws and the long–wavelength peculiarities of spatial correlators are caused by the balance between fluctuations of $z(r,t)$, generated by the random particles source, and the diffusive smoothing. It is important that in low dimensional systems ($d \leq 2$) with arbitrary relations between the rate constants the diffusive processes fail to equalize the spatial inhomogeneities. This leads to the spontaneous separation of homogeneous (in average) systems into the macroscopic domains containing particles of only one type. Besides, the mean concentrations increase with the generation time. These effects were observed numerically [32] and theoretically [33] on fractals. A similar picture can be observed in low dimensional systems with recombination, annihilation and multiplication which become separated into the domains with logarithmic law of growth [34].

The reaction kinetics drastically changes in systems with high volume fraction of reagents (above the percolation threshold), in polymer solutions and in crystals with topological defects. For these the correlation effects turn out to be decisive from the earliest times. First we consider the kinetics of (1b) for two percolation–like systems for which, in the absence of a reaction, particles A are localized in finite volumes. Particles A, each of a charge e, diffuse in presence of cartesian bias on a lattice, whose sites can be occupied by immobile, randomly distributed, entanglements S. It was shown [35] that if there exists a slow trapping of A by S the reaction kinetics goes on as follows. In absence of the electric field the long–time behavior is governed by the correlation–induced law (3). However, the intermediate kinetics, which define the annihilation of the bulk of A, is not mean–field too

$$\ln C_A(t) \approx -t^{d/d+1} . \tag{6}$$

In an electric field the decay (6) also defines the intermediate kinetics, but the long–time decay follows an exponential dependence upon time

$$\ln C_A(t) \approx -E \cdot t \quad \text{and} \quad \ln C_A(t) \approx -E^2 \cdot t$$

for $E \gg E_{\text{cr}}$ and $E \ll E_{\text{cr}}$, E_{cr} is some critical value of the external field. In the second case particles S are neutral with respect to the reaction and the decrease of A occurs in an encounter with particles of a third species T-trapping centers randomly distributed on a lattice among the sites which are not occupied by particles S. In the limit $t \to \infty$ $C_A(t)$ tends toward a finite limit $C_A(\infty)$, which is equal to the fraction of localization cavities which contain not a single trapping center. In the limit $t \to \infty$ we find [35]

$$\ln\{C_A(t) - C_A(\infty)\} \approx -t^{1/2}(E = 0) ,$$
$$\ln\{C_A(t) - C_A(\infty)\} \approx -\alpha \cdot \ln t (E > 0) .$$

Finally, we consider trapping kinetics in a special case when traps are attached to the segments of polymer chains, randomly distributed in viscous solvent. When the number of traps per each chain N is large, $N \gg 1$, the trapping kinetics exhibits unusual behavior at the intermediate times [36]

$$\ln C_A(t) \approx -t^{\gamma/\gamma+2} , \tag{7}$$

where γ exactly equals unity for $3d$ gaussian coils, $\gamma = d - 1; (d+2)/3$ for rod–like molecules and swollen coils respectively. At $t \to \infty$ limit the decay of (3) is restored. These results are in a good agreement with numerical data [37].

To summarize, we have presented some results corresponding to the many–particle kinetics of reactions in condensed media. We have shown that the kinetic behavior can be anomalous as a consequence of spatial correlations between particles but not due to the anomalous transport properties. Moreover, we have shown that the correlation–induced kinetics in some systems can govern the density evolution over the entire time domain.

References

1. H. Scher and E.W. Montroll: Phys. Rev. **B 12** 2455 (1975)
2. F.W. Schmidlin: Phys. Rev. **B 16** 2362 (1977)
3. S. Alexander et al.: Rev. Mod. Phys. **53** 175 (1981)
4. G.H. Weiss and R.J. Rubin: Adv. Chem. Phys. **52** 363 (1982)
5. J.W. Haus and K.W. Kehr: Phys. Rep. **150** 263 (1987)
6. H. Schnorer, D. Haarer and A. Blumen: Phys. Rev. **B 38** 8097 (1988)
7. M. Smoluchowsky: Z. Phys. **B 16** 321 (1915); **B 17** 557 (1916)
8. F.C. Collins and G.E. Kimball: J. Colloid. Sci. **4** 425 (1949)
9. S.F. Burlatsky et al.: Chem. Phys. Lett. **66** 565 (1979)
10. A.A. Ovchinnikov, S.F. Timashev and A.A. Belyy: *Kinetics of diffusion-controlled processes* (Nova Sci., N.Y. 1989)
11. D. Calef and J.M. Deutch: Ann. Rev. Phys. Chem. **34** 493 (1983)
12. P. Debye: Trans. Electrochem. Soc. **82** 265 (1942)
13. B.Y. Balagurov and V.T. Vaks: Sov. Phys. JETP **38** 968 (1974)
14. M.D. Donsker and S. Varadhan: Comm. Pure. Appl. Math. **28** 525 (1975)
15. A.A. Ovchinnikov and Y.B. Zeldovich: Chem. Phys. **28** 215 (1978)
16. R.F. Kayser and J.B. Hubbard: Phys. Rev. Lett. **51** 79 (1983)
17. S.F. Burlatsky and A.A. Ovchinnikov: Sov. Phys. JETP **65** 908 (1987)
18. I.M. Lifschitz: Sov. Phys. Usp. **7** 549 (1965)
19. S.F. Burlatsky: Teor. Exp. Chem. **14** 343 (1978)
20. D. Toussaint and F. Wilczek: J. Chem. Phys. **78** 2642 (1983)
21. K. Kang and S. Redner: Phys. Rev. Lett. **52** 955 (1984)
22. M. Bramson and J. Lebowitz: Phys. Rev. Lett. **61** 2397 (1988)
23. T. Ohtsuki: Phys. Lett. **A 106** 224 (1986)
24. Y.B. Zeldovich and A.A. Ovchinnikov: JETP Lett. **26** 440 (1977); [Sov. Phys. JETP **47** 829 (1978)]
25. K. Kang and S. Redner: Phys. Rev. **A 32** 437 (1985)
26. S.F. Burlatsky, A.A. Ovchinnikov and G.S. Oshanin: Sov. Phys. JETP **68** 1153 (1989)
27. G.S. Oshanin, A.A. Ovchinnikov and S.F. Burlatsky: J. Phys. **A 22** 977 (1989); **22** 947 (1989)
28. G.S. Oshanin: Sov. Chem. Phys. **9** 246 (1990)
29. G.S. Oshanin, S.F. Burlatsky and A.A. Ovchinnikov: Sov. Chem. Phys. **8** 372 (1989)
30. A.A. Ovchinnikov and S.F. Burlatsky: JETP Lett. **43** 638 (1986)
31. Y.C. Zhang: Phys. Rev. Lett. **59** 1726 (1987)
32. L.W. Anacker and R. Kopelman: Phys. Rev. Lett. **58** 289 (1987)
33. G.S. Oshanin, S.F. Burlatsky and A.A. Ovchinnikov: Phys. Lett. **A 139** 245 (1989)
34. S.F. Burlatsky and K.A. Pronin: J. Phys. **A 22** 346 (1989)
35. S.F. Burlatsky and A.A. Ovchinnikov: JETP Lett. **45** 567 (1987) S.F. Burlatsky and O.F. Ivanov: Sov. Phys. JETP **67** 1704 (1988)
36. S.F. Burlatsky and G.S. Oshanin: Phys. Lett.**A 145** 61 (1990)
37. G.S. Oshanin, A.V. Mogutov and S.F. Burlatsky: to be published

Fermionization of a Generalized Two-Dimensional Ising Model

A.I. Bugrij

Institute for Theoretical Physics, Metrologicheskaya 14, 252 130 Kiev, USSR

1. Introduction

The generalized two-dimensional Ising model is a system of spins that interact with partners of two neighboring coordination spheres. In the case of a square lattice the density of the Hamiltonian $h(\sigma_R)$ (the energy per plaquet) that satisfies the global Z_2-symmetry requirement includes not only pair interactions between first and second neighbors, but also four-spin interactions

$$h(\sigma_R) = -\sum_{i<l} J_{il}\sigma_R^i \sigma_R^l - J \prod_{i=1}^{4} \sigma_R^i, \quad i,l = 1,\ldots,4. \tag{1}$$

Here we use the following notations (see Fig. 1): $\sigma_R^i = \pm 1$ is the Ising spin; $R = (x,y)$ numerates the plaquet position in the lattice, $x = 1,\ldots,n$, $y = 1,\ldots,m$; the superscript i,l indicates at which vertex any given spin is situated; J_{il} are the parameters of pair interactions, J is the four-spin coupling constant.

We shall hence forward omit the index R for quantities referred to plaquets implied: for example, $\sigma^i = \sigma_R^i$, $\psi^i = \psi_R^i$. The total energy of a system

$$H\{\sigma\} = \sum_R h(\sigma_R)$$

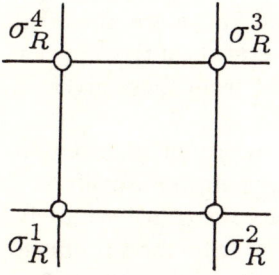

Fig. 1.

counts double the pair interactions that correspond to vertical and horizontal bonds $\sigma^1\sigma^4$, $\sigma^1\sigma^2$ (nearest neighbors), so the parameters J_{il} are related by

$$J_{12} + J_{34} = J_h, \quad J_{14} + J_{23} = J_v.$$

The parameters J_{13} and J_{24} correspond to the interaction of second neighbors along plaquet diagonals.

The partition function of the model can be represented as a sum of the product of statistical weights of the plaquets $z(\sigma_R)$

$$Z = \sum_{\{\sigma\}} \exp(-\beta H\{\sigma\}) = \sum_{\{\sigma\}} \prod_R z(\sigma_R) \quad (2)$$

where

$$z(\sigma_R) = \exp[-\beta h(\sigma_R)] = \exp(\sum_{i<l} K_{il}\sigma^i\sigma^l + K\prod_{i=1}^{4}\sigma^i)$$

$\beta = 1/T$ is the inverse temperature, $K = \beta J$, $K_{il} = \beta J_{il}$. To calculate the partition function (2) is a complicated problem, and its solution has been found only for some special cases:

i) $J_{13} = J_{24} = J = 0$ is an Ising model with nearest–neighbors interactions [1],
ii) $J_{13} = J = 0$ or $J_{24} = J = 0$ is an Ising model with nearest–neighbors interactions on a triangular lattice [2],
iii) $J_{12} + J_{34} = J_{14} + J_{23} = 0$ is an eight-vertex model without an external field [3].

As shown in [4], the system considered is generally equivalent to an eight-vertex model in an external field. An important, yet so far unsolved special case is the model involving a pair interaction of first and second neighbors and this corresponds to $J = 0$ in formula (1). It was discussed by many authors (see, for example, [5-9] and the literature quoted therein). The considerable efforts to study it were caused, on the one hand, by the intention to bring the properties of the model closer to real systems for most of which the restriction by an interaction of only nearest neighbors is a very rough approximation, and, on the other hand, by the desire to understand to what extent the conclusions concerning the critical properties of spin systems that follow from the exact solutions discussed above are universal. More specifically, there is so far no satisfactory interpretation of the fact that the critical exponents of models i) and ii) are different from those given by model iii).

Among the numerous procedures for obtaining or reproducing exact solutions of two-dimensional spin models we can distinguish, tentatively in a measure of course, the three main approaches, these are: the transfer-matrix, the combinatorial, and fermionization methods. The later method produces the results in a very economical and elegant manner [10]. The method of an auxiliary Grassmann field proposed in [11] and developed in [12-14] was an important step in refining the procedure to fermionize spin lattice systems. Within this approach, it is easy enough to obtain solutions to corresponding problems on lattices of finite sizes with free boundaries, but this not quite so simple in the transfer–matrix or combinatorial approaches. It was just the method of an auxiliary Grassmann field that enabled for the first time a solution to such a problem [15].

It is to be noted that the technique of Grassmann variables was first used in [16-17] within the combinatorial approach to Ising problem. In contrast to this, the authors of [11] found a direct (local) correspondence between the initial set of spin variables and the Grassmann field. Technically the central point here is the factorization of coupled terms in the partition function using integrals over Grassmann variables such as

$$\exp(K\sigma\sigma') = \int d\varphi d\varphi' \exp(\varphi\varphi' \text{th} K)(\sigma + \varphi)(\sigma' + \varphi') \qquad (3)$$

and this makes it possible to sum over the spins independently for every lattice site. But the idea was difficult to realize: the noncommutativity of the terms in (3) linear in Grassmann variables hinders their reordering. To by-pass this obstacle it was necessary to introduce, along with the Grassmannian field, other additional anti-commutative quantities later excluded from the ultimate expressions. For example, in [11] Clifford constants are used, while in [12,13] a set of fermion creation and annihilation operators are employed that correspond to various lattice sites. The elaboration of the method [13] enabled an accomplished mathematical formulation in terms of fermion operator algebra and gave an insight into the physical contents of the procedure of fermionizing spin–lattice systems. Meanwhile, one of the major advantages of the method of an auxiliary Grassmann field – its strictness and simplicity – was concealed to en extent in these approaches. And the further refinement of the method was aimed at avoiding any more mathematical objects, in addition to Grassmannian ones, that are introduced for purely technical purposes. In [14], this was accomplished by quite an innovative "mirror" ordering of the initial expressions, and in [12] by means of two sets of Grassmannian variables.

2. Transformation of the partition function to a functional integral

The partition function was transformed to the functional Grassmann integral [11-15,18] in the Ising model with nearest-neighbor interactions only. In the present paper we propose a procedure to generalize the method of an auxiliary Grassmann field to Ising systems in which not only nearest neighbors interact. The main feature of the method proposed in to introduce two Grassmann fields φ and χ that commute with each other, and this makes unnecessary to employ any additional anticommuting quantities and allows us to easily formalize the fermionization procedure. The system that we obtain by summing over spin degrees of freedom can be interpreted as a system composed of fermions of two different kinds. But next we show that such a system is equivalent to a system composed of fermions of one kind. The conclusion is based on the following relation between the Grassmann functional integrals (see Appendix)

$$\int D\varphi D\chi \exp[-\frac{1}{2}(\chi, A\chi) + \frac{1}{2}(\varphi, B\varphi) + (\varphi, C\chi)] =$$

$$= \int D\varphi D\psi \exp[\frac{1}{2}(\psi, A\psi) + \frac{1}{2}(\varphi, B\varphi) + (\varphi, C\varphi)]$$

where $[\varphi, \chi] = 0$, $\{\varphi, \psi\} = 0$. Restricting ourselves to a minimum set of auxiliary fields – one component for every plaquet spin – we write down the following representation

$$\prod_R z(\sigma_R) = (-c)^{mn} \int D\varphi D\chi \exp[\sum_R \mathcal{L}_R^1(\varphi, \chi)] \prod_R \omega_R(\varphi, \chi; \sigma) \quad (4)$$

where $\varphi = (\varphi^1, \varphi^2)$, $\chi = (\chi^1, \chi^2)$ are Grassmannian

$$\{\varphi_R^i, \varphi_{R'}^l\} = 0, \quad \{\chi_R^i, \chi_{R'}^l\} = 0, \quad [\varphi_R^i, \chi_{R'}^l] = 0; \quad (5)$$

$$D\varphi = \prod_R d\varphi_R^1 d\varphi_R^2, \quad D\chi_R = \prod_R d\chi_R^1 d\chi_R^2. \quad (6)$$

The order of co-factors in (6) that give the integration measure is unimportant, because its change can result only in a changed common sign before the integral that can easily be reconstructed from the positive definiteness of the partition function. The dependence of the "Lagrangian" $\mathcal{L}_R^1(\varphi, \chi)$ and the functions $\omega_R(\varphi, \chi; \sigma)$ on the fields in (4) is localized in the plaquet, or else we could not be able to carry out the integration. Obviously, the most general form of the Lagrangian is then the following

$$\mathcal{L}_R^1(\varphi, \chi) = -\frac{1}{2}(\chi, A\chi) + \frac{1}{2}(\varphi, B\varphi) + (\chi, C\varphi) - g\varphi^1\varphi^2\chi^1\chi^2. \quad (7)$$

The matrix A, B and C that define the quadratic form in (7) contains 6 arbitrary parameters which, along with the "coupling constant" g and the general normalization coefficient c, should be defined through the initial parameters K_{il}, K, involved in the plaquet statistical weights $z(\sigma_R)$. Taking into account the commutational properties of the fields φ, χ (5) it is seen that

$$[\mathcal{L}_R^1(\varphi, \chi), \mathcal{L}_{R'}^1(\varphi, \chi)].$$

Since the terms $\mathcal{L}_R^1$ belonging to different plaquets are commutative, the multiple integral (4) is reduced to an iterated one, and the integration is

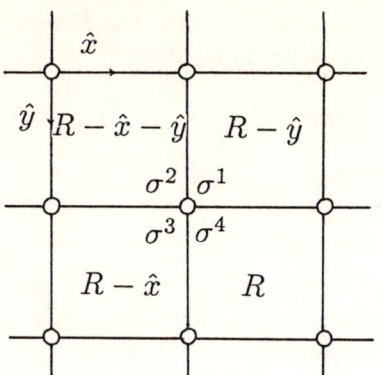

Fig. 2.

carried out independently for every plaquet, even if the functions $w_R(\varphi, \chi; \sigma)$ do not commute with each other $[w_R(\varphi, \chi; \sigma), w_{R'}(\varphi, \chi; \sigma)] \neq 0$. As a result, we obtain for the plaquet statistical weight

$$z(\sigma_R) = -c \int d\varphi^1 d\varphi^2 d\chi^1 d\chi^2 \exp[-\frac{1}{2}(\chi, A\chi) + \frac{1}{2}(\varphi, B\varphi) + (\varphi, C\chi) - g\varphi^1\varphi^2\chi^1\chi^2] w_R(\varphi, \chi; \sigma) \,. \tag{8}$$

So we are now left with the function $w_R(\varphi, \chi; \sigma)$ whose form is practically unambiguously defined from the following two conditions: first, the possibility to localize the spin dependence in the integrand (4), and, second, the expression that we have after summing over the spin should be represented as an exponent of the quadratic form. Before writing down explicitly relevant restrictions we note that any given spin is involved simultaneously in four neighboring plaquets, because in the notation adopted

$$\sigma^2_{R-\hat{x}-\hat{y}} = \sigma^1_{R-\hat{x}} = \sigma^3_{R-\hat{x}} = \sigma^4_R$$

where $\hat{x}$ and $\hat{y}$ denote the unit vectors along horizontal and vertical axes, respectively (see Fig. 2).

So long as we shall be concerned with noncommuting quantities, it is necessary to adopt a certain ordering of the product $\prod_R z(\sigma_R)$, for example in rows

$$\prod_R z(\sigma_R) = \prod_{y=1}^{m} \prod_{x=1}^{n} z(\sigma_{x,y})$$
$$= \underbrace{(z_{1,1} z_{2,1} \ldots z_{n,1})}_{1\text{-st row}} \underbrace{(z_{1,2} z_{2,2} \ldots z_{n,2})}_{2\text{-nd row}} \ldots \underbrace{(z_{1,m} z_{2,m} \ldots z_{n,m})}_{m\text{-th row}} \,. \tag{9}$$

Taking the above reasoning into account we require that the function $\omega_R(\varphi, \chi; \sigma)$ should have a factorized dependence on the spins

$$\omega_R(\varphi, \chi; \sigma) = v_R^1(\varphi, \chi; \sigma^1) v_R^2(\varphi, \chi; \sigma^2) v_R^4(\varphi, \chi; \sigma^4) v_R^3(\varphi, \chi; \sigma^3) \qquad (10)$$

in order to regroup the cofactors in (9) and obtain

$$\prod_R \omega_R = \prod_R v_R^1 v_R^2 v_R^4 v_R^3 \longrightarrow \prod_R v_{R-\hat{x}-\hat{y}}^2 v_{R-\hat{y}}^1 v_{R-\hat{x}}^3 v_R^4$$

(the right-hand side of this relation has an expression dependent only on the spin σ_R^4). But for such a reordering to be possible the functions v_R^1, v_R^2 should commute with v_R^3, v_R^4 that belong to different R:

$$[v_R^j(\varphi, \chi; \sigma^j), v_{R'}^k(\varphi, \chi; \sigma^k)] = 0 \ ; \quad j = 1, 2 \ , \quad k = 3, 4 \ . \qquad (11)$$

We note that giving ω_R as a product ordered according to (10) requires no more permutations, except (11), because the spins of the row have already been ordered in increasing coordinate X. Taking into account the fact that the functions $v_R^i(\varphi, \chi; \sigma^i)$ are decomposed into a polynomial linear in σ_R^i and not higher that the fourth degree in Grassmann variables φ_R^i and χ_R^i, as well as equation (11) and the condition for $\sum_{\{\sigma\}} \prod_R \omega_R$ to be Gaussian

$$\frac{1}{2} \sum_{\sigma = \pm 1} v_{R-\hat{x}-\hat{y}}^2 v_{R-\hat{y}}^1 v_{R-\hat{x}}^3 v_R^4 = \exp[\mathcal{L}_R^2(\varphi, \chi)] \ , \qquad (12)$$

where $\mathcal{L}_R^2(\varphi, \chi)$ is the quadratic function of φ and χ, we get an explicit form of the functions $v_R^i(\varphi, \chi; \sigma^i)$ up to unimportant scale factors before φ and χ

$$v^1 = (\sigma^1 + \chi^1) \ , \quad v^2 = (\sigma^2 + \chi^2) \ ,$$
$$v^3 = (\sigma^3 + \varphi^1) \ , \quad v^4 = (\sigma^4 + \varphi^2) \ , \qquad (13)$$
$$\omega_R(\varphi, \chi; \sigma) = (\sigma^1 + \chi^1)(\sigma^2 + \chi^2)(\sigma^4 + \varphi^2)(\sigma^3 + \varphi^1) \ .$$

Substituting (13) into (12) and summing over σ, we get

$$\mathcal{L}_R^2(\varphi, \chi) = \ln\left[\frac{1}{2} \sum_{\sigma = \pm 1} (\sigma + \chi_{R-\hat{x}-\hat{y}}^2)(\sigma + \chi_{R-\hat{y}}^1)(\sigma + \varphi_{R-\hat{x}}^1)(\sigma + \varphi_R^2)\right]$$
$$= -\chi_{R-\hat{y}}^1 \chi_{R-\hat{x}-\hat{y}}^2 + \varphi_{R-\hat{x}}^1 \varphi_R^2 + \chi_{R-\hat{y}}^1 \varphi_{R-\hat{x}}^1 + \chi_{R-\hat{y}}^1 \varphi_R^2 +$$
$$+ \chi_{R-\hat{x}-\hat{y}}^2 \varphi_{R-\hat{x}}^1 + \chi_{R-\hat{x}-\hat{y}}^2 \varphi_R^2 \ .$$

All the terms of $\mathcal{L}_R^1(\varphi, \chi)$ and $\mathcal{L}_R^2(\varphi, \chi)$ commute with one another and terms belonging to other plaquets, so that on summing (4) over all lattice spins we arrive at the following representation for the partition function

$$Z = 2^{(m+1)(n+1)}(-c)^{mn} \int D\varphi \, D\chi \exp\left\{\sum_R [\mathcal{L}_R^1(\varphi,\chi) + \mathcal{L}_R^2(\varphi,\chi)]\right\}.$$

Because the Lagrangian $\mathcal{L}_R^2(\varphi,\chi)$ is summed in R, it is more convenient to represent this in the following compact form

$$\mathcal{L}_R^2 = \varphi^1 \nabla_x \varphi^2 - \chi^1 \nabla_{-x} \chi^2 + \varphi^1(\nabla_{x,-y}\chi^2 + \nabla_{-y}\chi^2) + \varphi^2(\nabla_{-y}\chi^1 + \nabla_{-x,-y}\chi^2),$$

$\nabla_{-x,-y}$ denotes $\nabla_{-x}\nabla_{-y}$, and $\nabla_{\pm x}$, $\nabla_{\pm y}$ the operators of one-step shift of a lattice along horizontal or vertical directions, respectively,

$$\nabla_{\pm x}\varphi_R^i = \varphi_{R\pm\hat{x}}^i, \qquad \nabla_{\pm x}\chi_R^i = \chi_{R\pm\hat{x}}^i,$$
$$\nabla_{\pm y}\varphi_R^i = \varphi_{R\pm\hat{y}}^i, \qquad \nabla_{\pm y}\chi_R^i = \chi_{R\pm\hat{y}}^i.$$

Their matrix elements are expressed through Kronecker δ–symbols

$$(\nabla_{\pm x})_{RR'} = \delta(y - y')\delta(x - x' \pm 1),$$
$$(\nabla_{\pm y})_{RR'} = \delta(x - x')\delta(y - y' \pm 1).$$

The Lagrangian $\mathcal{L}_R^2(\varphi,\chi)$ can be written as a scalar product in a way similar to (7)

$$\mathcal{L}_R^2(\varphi,\chi) = -\frac{1}{2}(\chi, A'x) + \frac{1}{2}(\varphi, B'\varphi) + (\varphi, C'x), \qquad (14)$$

where

$$A' = \begin{pmatrix} 0 & \nabla_{-x} \\ -\nabla_x & 0 \end{pmatrix}, \quad B' = \begin{pmatrix} 0 & \nabla_x \\ -\nabla_{-x} & 0 \end{pmatrix}, \quad C' = \begin{pmatrix} \nabla_x \nabla_{-y} & \nabla_{-y} \\ \nabla_{-y} & \nabla_{-x}\nabla_{-y} \end{pmatrix}.$$

It is now easy to pass from the field χ that commutes with φ to the field ψ that anticommutes with φ

$$\{\psi_R^i, \psi_{R'}^l\} = 0, \quad \{\psi_R^i, \varphi_{R'}^l\} = 0.$$

According to the result given in the Appendix (formula (A14), (A15), (A16)) the replacement of the field χ by ψ results only in a changed sign for the matrices A, A' and for the quartic term in the Lagrangian $\mathcal{L}_R^1$, i.e. instead of (7), (14) we get

$$\mathcal{L}_R^1(\varphi,\psi) = \frac{1}{2}(\psi, A\psi) + \frac{1}{2}(\varphi, B\varphi) + (\varphi, C\psi) + g\varphi^1\varphi^2\psi^1\psi^2,$$
$$\mathcal{L}_R^2(\varphi,\psi) = \frac{1}{2}(\psi, A'\psi) + \frac{1}{2}(\varphi, B'\varphi) + (\varphi, C'\psi). \qquad (15)$$

In a four-component form

$$\psi_R = (\psi_R^1, \psi_R^2, \varphi_R^1, \varphi_R^2),$$
$$\mathcal{L}_R(\psi) = \mathcal{L}_R^1(\varphi,\psi) + \mathcal{L}_R^2(\varphi,\psi) = \frac{1}{2}(\psi, D\psi) + \frac{1}{8}g(\psi, \Gamma\psi)^2,$$

where

$$D = \begin{pmatrix} A+A' & -(C+C') \\ C+C' & B+B' \end{pmatrix} \quad \Gamma = \begin{pmatrix} \tau & 0 \\ 0 & \tau \end{pmatrix} \quad \tau = \begin{pmatrix} 0 & 1 \\ -1 & 0 \end{pmatrix}.$$

3. Relationship between the representation parameters

We now turn back to the relationship between the parameters of the Lagrangian (7) and the parameters of the initial model. It is easy to see that the plaquet statistical weight $z(\sigma_R)$ can be expanded in a polynomial of independent spin combinations

$$z(\sigma_R) = c(1 + \sum_{i<l} a_{il}\sigma^i\sigma^l + a \prod_{i=1}^{4} \sigma^i) . \tag{16}$$

Introducing the notation $\langle r(\sigma) \rangle$, for the plaquet average of the value $r(\sigma)$

$$\langle r(\sigma) \rangle = \sum_{\sigma_1,\ldots,\sigma_4} r(\sigma)z(\sigma) \Big/ \sum_{\sigma_1,\ldots,\sigma_4} z(\sigma)$$

we find that the coefficients (16) a_{il}, a and c are in one-to-one correspondence with the parameters of the Hamiltonian (1) J_{il}, J

$$a_{il} = \langle \sigma^i\sigma^l \rangle , \qquad a = \langle \sigma^1\sigma^2\sigma^3\sigma^4 \rangle , \qquad c = 2^{-4} \sum_{\sigma_1,\ldots,\sigma_4} z(\sigma) . \tag{17}$$

If we use the following notation for the elements of the matrices A, B and C

$$A = \begin{pmatrix} 0 & a_{12} \\ -a_{12} & 0 \end{pmatrix} \quad B = \begin{pmatrix} 0 & a_{34} \\ -a_{34} & 0 \end{pmatrix} \quad C = \begin{pmatrix} -a_{13} & a_{14} \\ a_{23} & -a_{24} \end{pmatrix}$$

the integration result in (8) will be the same as the right-hand side of (16), and as the coefficient for $\sigma^1\sigma^2\sigma^3\sigma^4$ in (16) we have

$$a = g + a_{12}a_{34} - a_{13}a_{24} + a_{14}a_{23}$$

and this yields for the four-fermion coupling constant:

$$g = \langle \sigma^1\sigma^2\sigma^3\sigma^4 \rangle - \langle \sigma^1\sigma^2 \rangle\langle \sigma^3\sigma^4 \rangle + \langle \sigma^1\sigma^3 \rangle\langle \sigma^6\sigma^4 \rangle - \langle \sigma^1\sigma^4 \rangle\langle \sigma^2\sigma^3 \rangle . \tag{18}$$

The partition function of a generalized Ising model thus has the following four-component form:

$$Z = 2^{(m+1)(n+1)} c^{mn} \int D\psi \exp\left\{ \frac{1}{2} \sum_R [(\psi, D\psi) + \frac{1}{4}g(\psi, \Gamma\psi)^2] \right\} \tag{19}$$

where the constants c, g are given by formulas (17), (18). The matrix D - a discrete analogy of the Dirac operator - and the matrix Γ have the following form

$$D = \begin{pmatrix} 0 & a_{12} - \nabla_{-x} & -a_{13} + \nabla_{-x}\nabla_y & a_{14} + \nabla_y \\ -a_{12} + \nabla_x & 0 & a_{23} + \nabla_y & -a_{24} + \nabla_x \nabla_y \\ a_{13} - \nabla_x \nabla_{-y} & -a_{23} - \nabla_{-y} & 0 & a_{34} - \nabla_x \\ -a_{14} - \nabla_{-y} & a_{24} - \nabla_{-x}\nabla_y & -a_{34} + \nabla_{-x} & 0 \end{pmatrix} \qquad (20)$$

$$\Gamma = \begin{pmatrix} 0 & 1 & 0 & 0 \\ -1 & 0 & 0 & 0 \\ 0 & 0 & 0 & 1 \\ 0 & 0 & -1 & 0 \end{pmatrix}.$$

The presence of a quartic term does not allow us to integrate (19) explicitly in the general case. Let us now establish for which set of the parameters of the Hamiltonian (1) the coupling constant g vanishes, that is, conditions under which the generalized Ising model is equivalent to a free fermion system. By substituting in (18) the explicit expressions for the plaquet averages $\langle \sigma^i \sigma^l \rangle$ and $\langle \sigma^1 \sigma^2 \sigma^3 \sigma^4 \rangle$, we obtain

$$g = (16c)^{-2}(\text{th}2K + \text{th}2K_{13} \cdot \text{th}2K_{24}) . \qquad (21)$$

The vanishing of (21) fixes a surface in the space of parameters K_{il}, K on which the free energy is explicitly expressed through the determinant of operator D.

$$F = \ln Z = \frac{1}{2}(\det D) + mn \ln c + (m+1)(n+1)\ln 2 . \qquad (22)$$

As regards the thermodynamic characteristic of the system, they are determined by the dependence of a free energy on K_{il}, K along the ray

$$K/J = K_{il}/J_{il} = \beta.$$

Therefore the thermodynamic quantities can be obtained from (22) with arbitrary and fixed parameters of the Hamiltonian J_{il}, J only for some special temperature values that correspond to the real and positive roots of the equation

$$\text{th}(2\beta J) + \text{th}(2\beta J_{13}) \cdot \text{th}(2\beta J_{24}) = 0 . \qquad (23)$$

4. Ising model with pair interactions of nearest– and next–nearest neighbors

As seen from (21), the four-fermion coupling constant is identically equal to zero at any temperatures if the following conditions are valid

$$J = J_{13} = J_{24} = 0 \quad \text{or} \quad J = J_{13} = 0 \quad \text{or} \quad J = J_{24} = 0 . \qquad (24)$$

The conditions (24), as mentioned above, are met in the Ising model with nearest-neighbors interactions on rectangular and triangular lattices. Also, another interesting version on the spin system (19) is realized under such

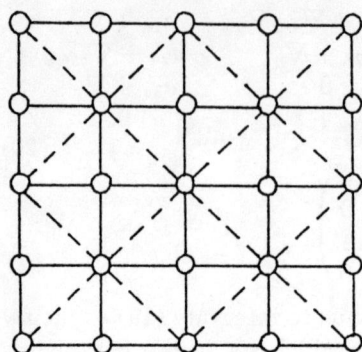

Fig. 3.

conditions. This is the case when the second and the third of the conditions (24) are satisfied for different plaquets alternating vertically and horizontally: a corresponding situation is shown in Fig. 3.

Here we deal with a model in which half of the spin interact with the first and second neighbors, and the second half only with the nearest neighbors. One can assume that the thermodynamics of such a model will not be different qualitatively from a model with nearest and next-nearest neighbor interaction when only $J = 0$ and the four-fermion coupling constant in (19) does not vanish

$$g = \text{th}^2 2K_2/(e^{2K_2}\coth 2K_1 + 2 + e^{-2K_2})^2 \tag{25}$$

here we set $K_1 = \beta J_1$, $K_2 = \beta J_2$,

$$J_{12} = J_{34} = J_{14} = J_{23} = J_1/2 , \quad J_{13} = J_{24} = J_2 .$$

Because $g \sim \text{th}^2 2K_2$, calculating (19) in terms of perturbation theory will correspond to some rearranged high-temperature expansion in $\text{th}^2 2K_2$ contrary to ordinary high-temperature series [5] where the expansion is in the powers of $\text{th} K_2$. It is of interest that even in the zeroth (in g) order of perturbation theory the phase picture of the model has a form very close to that expected from Monte-Carlo simulation [20] and various approximate calculations [5-7]. Calculating the determinant of the operator (20), we find in the limit $m, n \to \infty$

$$Sp(\ln D)/mn = \frac{1}{4\pi^2} \int_{-\pi}^{\pi} dx\,dy \ln[\alpha - \gamma(\cos x + \cos y) + \delta \cos x \cdot \cos y] ,$$

where the parameters α, γ and δ are expressed through K_1 and K_2 in the following way

$$\alpha - 2\gamma + \delta = [2(1-a)^2 - (1+b)^2]^2$$
$$\alpha + 2\gamma + \delta = [2(1+a)^2 - (1+b)^2]^2 \tag{26}$$
$$\alpha - \delta = [2(1+a^2) - (1-b)^2]^2$$

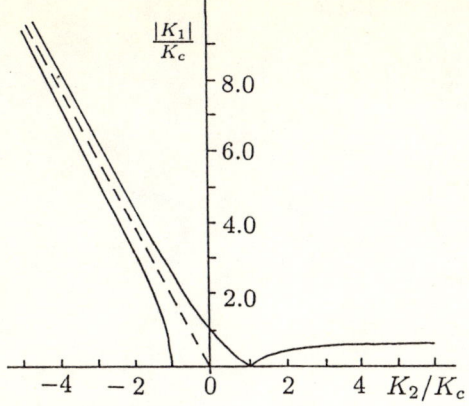

Fig. 4.

where
$$a = \sh 2K_1/(\ch 2K_1 + 2e^{-2K_2} + e^{-4K_2}),$$
$$b = (\ch 2K_1 - e^{-4K_2})/(\ch 2K_1 + 2e^{-2K_2} + e^{4K_2}).$$

The critical curve on a plane (K_1, K_2) is determined by the zero modes of the operator D, and this corresponds to the right-hand sides of the equations (26) becoming zero. The curve is plotted in Fig. 4. Typically, the critical curve goes through all the "control points": $K_1 = \pm K_c$ at $K_2 = 0$ and $K_2 = \pm K_c$ at $K_1 = 0$, where $K_c = \frac{1}{2}\ln(\sqrt{2}+1)$ is the critical point of a two-dimensional Ising model with nearest-neighbor interaction. If for $K_2 = 0$ the above result is trivial, because the constant g in virtue of (25) becomes zero and the approximation under consideration becomes exact, in the case $K_1 = 0$ the parameter $g \neq 0$ and a corresponding result comes as a pleasant surprise. Moreover, not only the critical point, but also the free energy determined by formula (22) coincides with the exact expression

$$f = \lim_{m,n\to\infty} (\ln z/mn) = \ln 2 +$$
$$+ \frac{1}{8\pi^2} \int_{-\pi}^{\pi} dx dy (\cosh^2 2K_2 - 2\sinh 2K_2 \cos x \cdot \cos y)$$

in spite of the fact that the four-fermion coupling parameter (25) is different from zero. One more feature of the approximation under consideration is not worthy. The analysis of the symmetry of the ground state of the Hamiltonian of the system considered has shown that there is no phase transition when the parameters of the Hamiltonian are related as follows [6]

$$2J_2 = -|J_1|$$

Fig. 4 shows that the effect is also reproduced here: the asymptotes of the plot branches $|K_1| = f(K_2)$ have a slope $|K_1|/K_2 = -2$ and cross the K_2 axis at points

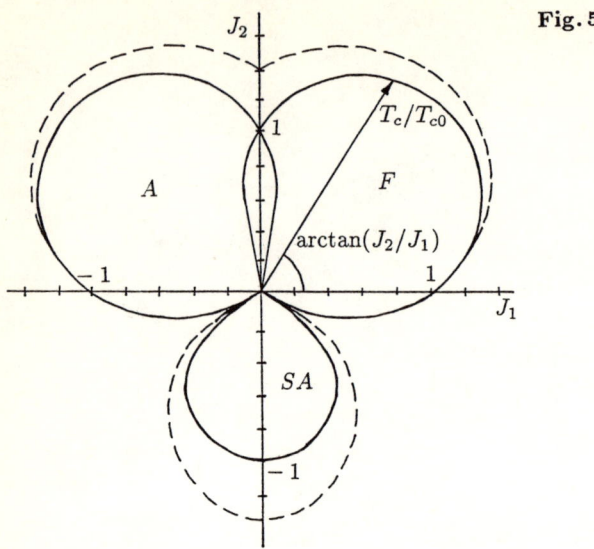

Fig. 5

$$K_2 = \frac{1}{4}\ln(\sqrt{3}-1), \quad K_2 = \frac{1}{8}\ln 2.$$

The phase diagram in polar coordinates $T_c/T_{co} = f(\omega)$, $\text{tg}\omega = J_1/J_2$ (where T_c is the phase transition temperature for the case when the first and second neighbors interact, and T_{co} is that for the case when only the nearest neighbors interact) is given in Fig. 5. In the same figure, for comparison, the dashed line shows the result obtained within the fermion representation framework in [6]. Three types of orderings are known to be possible in the model with the first and second neighbors interacting, depending upon the values of the parameters J_1 and J_2: ferromagnetic, antiferromagnetic and superantiferromagnetic types. The regions of different orderings denoted, respectively, by F, A and SA are shown in Fig.5, and as is seen from this figure, the superantiferromagnetic region is separated from the ferromagnetic and antiferromagnetic ones in full agreement with the symmetry arguments. We also note that there is a range of J_1 and J_2 at which the system undergoes three phase transitions, just as in the exactly solvable case [19] with a lattice, Fig.3 (the authors of that paper find an analogy of this behaviour in ferroelectrics).

Conclusion

The problem of fermionizing the Ising model with a pair interaction of first and second neighbors was earlier considered within the combinatorial approach [6,7]. The representations derived in those papers are different from each other: one of them [6] has the advantage over the other [7] in that in the Gaussian approximation it has a branch that corresponds to a superantiferromagnetic phase, and the disadvantage that the value of the critical points for the exactly solvable case $K_1 = 0$ is incorrect. The result of the present paper is different from the result of [6,7] both in the method and in the form of the final expressions, combining the advantages of both. The great possibilities of the method of an auxiliary Grassmann field are shown by the simple fact that the derivation of the fermion representation occupies only a few papers in contrast to the very cumbersome and strictly specialized methods employed previously to solve the problem. The specific representation obtained is quite informative as it follows from the application to the analysis of Ising model with first and second neighbors interactions: the basic features of the model are correctly reproduced both qualitatively and quantitatively already in the zeroth order in g. It appears, therefore, quite promising to use not only perturbation theory, but also other schemes such as a Hartree-Fock, variational, renormalization group methods, to refine corresponding results.

The availability of three different fermion representations for the same model indicates that the transition from a spin to a fermion system is ambiguous. If we could describe a total class of fermion representations that correspond to a given spin system, we could probably find among them representations which admit solutions without the restriction (23). We note that the condition (23) is equivalent to that of the "free fermion model" [21], and both the representations on a relevant hypersurface in the space of parameters lead to the same result for a free energy.

For the sake of brevity we have not covered some noteworthy points, in particular that of taking boundary conditions into account. We only mention that the presence of boundaries modifies corresponding contributions to the Lagrangian (15). Within the fermionization scheme considered the explicit form of these terms can easily be derived for free and periodic boundary conditions in a manner similar to that used in [15,18] for the model with nearest-neighbor interaction.

The author extends his gratitude to D.N.Zubarev, V.N.Popov, P.I.Fomin, V.N.Shadura for valuable comments and helpful discussions and also thanks to V.N.Plechko for communicating his result on the fermionization problem.

Appendix

We now find an explicit form of the generating functional $Q_c\{\varphi\}$ given as a continuous integral in Grassmannian variables with Gauss distribution

$$Q_c\{\varphi\} = \int D\chi \exp[\frac{1}{2}(\chi, A\chi) + (\chi, \varphi)] \qquad (A1)$$

in the case when the external Grassmann field φ is commutative with a field χ over which the integration is performed:

$$\{\varphi_k, \varphi_l\} = 0, \qquad \{\chi_k, \chi_l\} = 0, \qquad [\varphi_k, \chi_l] = 0;$$

$$D\chi = \prod_{k=1}^{N} d\chi_k, \qquad k, l = 1, 2, \ldots, N, \qquad N \text{ is even}.$$

The scalar products are defined ordinarily

$$(\chi, \varphi) = \sum_k \chi_k \varphi_k, \qquad (\chi, A\chi) = \sum_{k,l} \chi_k \chi_l A_{kl}.$$

The matrix A is antisymmetric ($A = -A^T$), because the Grassmann variables are anticommutative. If the external field φ anticommutes with the internal one ψ

$$\{\varphi_k, \psi_l\} = 0, \qquad \{\psi_k, \psi_l\} = 0$$

the result is available

$$Q_a\{\varphi\} = \int D\psi \exp[\frac{1}{2}(\psi, A\psi) + (\psi, \varphi)] = \det^{1/2} A \cdot \exp[\frac{1}{2}(\varphi, A^{-1}\varphi)]. \quad (A2)$$

Formula (A2) is easily proven by shifting the integration variable

$$\psi \to \psi' = \psi - A^{-1}\varphi.$$

This technique, however, is not applicable to the case under consideration, because having been shifted to some Grassmann constant ξ that commutes with χ

$$\chi \to \chi' = \chi - \xi$$

the transformed quantity χ' looses its anticommutational property

$$\{\chi'_k, \chi'_l\} = -2\{\chi_k, \xi_l\} \neq 0.$$

Yet, the integration in (A1) can be performed explicitly for an arbitrary N using the property of antisymmetric matrices that are reduced to a quasi-diagonal representation by means of an orthogonal real transformation O

$$\chi \to \chi' = O\chi, \qquad OO^T = 1;$$

$$A \to A' = OAO^T = \begin{pmatrix} \lambda_1 \tau & 0 & \cdots & 0 \\ 0 & \lambda_2 \tau & \cdots & 0 \\ \vdots & \vdots & \ddots & \vdots \\ 0 & 0 & \cdots & \lambda_{N/2} \tau \end{pmatrix} \quad (A3)$$

where τ is a 2×2 matrix

$$\tau = \begin{pmatrix} 0 & 1 \\ -1 & 0 \end{pmatrix},$$

the numbers λ_k are related to the eigenvalues μ_k of the matrix A by

$$\mu_{2k-1} = i\lambda_k, \qquad \mu_{2k} = -i\lambda_k.$$

The replacement of variables (A3) is admissible, because

$$\{\chi'_k, \chi'_l\} = 0.$$

Since the Berezian of transformation (A3) is equal to one, the integral (A2) can be written as

$$Q_c\{\varphi\} = \int D\chi \exp[\tfrac{1}{2}(\chi, A'\chi) + (\chi, \varphi')], \qquad (A4)$$

where $\varphi' = O\varphi$. The integration in (A4) is now split into a product of double integrals

$$Q_c\{\phi\} = \prod_{k=1}^{N/2} \int d\chi_{2k-1} d\chi_{2k} \exp(\lambda_k \chi_{2k-1}\chi_{2k} + \chi_{2k-1}\varphi'_{2k-1} + \chi_{2k}\varphi'_{2k}) \qquad (A5)$$

as a result of the block-diagonal structure of the matrix A' (A3). We perform the integration in (A5), obtaining

$$Q_c\{\varphi\} = \prod_{k=1}^{N/2}(\lambda_k + \varphi'_{2k-1}\varphi'_{2k}) = \left(\prod_{k=1}^{N/2} \lambda_k\right) \exp\left(\sum_{k=1}^{N/2} \lambda_k^{-1} \varphi'_{2k-1}\varphi'_{2k}\right). \qquad (A6)$$

The sum in the exponent on the right-hand side of (A6) is just

$$\sum_{k=1}^{N/2} \lambda_k^{-1} \varphi'_{2k-1}\varphi'_{2k} = -\tfrac{1}{2}(\varphi', A'^{-1}\varphi') = -\tfrac{1}{2}(\varphi, A^{-1}\varphi) \qquad (A7)$$

and the product

$$\prod_{k=1}^{N/2} \lambda_k = \det{}^{1/2}(A') = \det{}^{1/2}(A) \qquad (A8)$$

Substituting (A7) and (A8) into (A6), we obtain

$$Q_c\{\varphi\} = \int D\chi \exp\left[\tfrac{1}{2}(\chi, A\chi) + (\chi, \varphi)\right] = \det{}^{1/2} A \cdot \exp\left[-\tfrac{1}{2}(\varphi, A^{-1}\varphi)\right] \qquad (A9)$$

We can see that this expression is different from the expression (A2) for the ordinary (anticommutative) case only in a sign before the quadratic form in the exponent.

An obvious consequence of (A9) is the equality of the integrals

$$Z_a = \int D\psi \, D\varphi \exp\left[\tfrac{1}{2}(\psi, A\psi) + \tfrac{1}{2}(\varphi, B\varphi) + (\psi, C\varphi)\right] \qquad (A10)$$

$$Z_c = \int D\chi D\varphi \exp\left[\frac{1}{2}(\chi, A\chi) - \frac{1}{2}(\varphi, B\varphi) + (\chi, C\varphi)\right] \qquad (A11)$$

Integrating in (A10) over φ, and then over ψ (in (A11) over φ and over χ, respectively), we obtain

$$Z_a = Z_c = \det{}^{1/2}(B)\det{}^{1/2}(A + CB^{-1}C^T) = \det{}^{1/2}\begin{pmatrix} A & C \\ -C^T & B \end{pmatrix} \qquad (A12)$$

where the last in the chain of equalities is the result of applying Sure's formula [22] for block–matrix determinants.

It is easy to prove similar equalities when the exponents (A10), (A11) involve quartic terms such as

$$\sum_k (\varphi_k, \Gamma\varphi_k)(\chi_k, \Gamma\chi_k) \,, \quad \sum_k (\varphi_k, \Gamma\varphi_k)(\psi_k, \Gamma\psi_k) \qquad (A13)$$

where the matrix Γ acts upon the inner indices of fields φ, χ. Specifically, if the fields are two-component

$$\varphi_k = (\varphi_k^1, \varphi_k^2)\,, \quad \chi_k = (\chi_k^1, \chi_k^2)\,, \quad \psi_k = (\psi_k^1, \psi_k^2)$$

the term (A13) reduces to a single combination

$$\sum_k \varphi_k^1 \varphi_k^2 \chi_k^1 \chi_k^2 \,, \quad \sum_k \varphi_k^1 \varphi_k^2 \psi_k^1 \psi_k^2$$

Let

$$Z_a = \int D\varphi D\psi \exp[\frac{1}{2}(\psi, A\psi) + \frac{1}{2}(\varphi, B\varphi) + (\psi, C\varphi) \\ + \frac{g}{4}\sum_k (\varphi_k, \Gamma\varphi_k)(\psi_k, \Gamma\psi_k)] \,, \qquad (A14)$$

$$Z_c = \int D\varphi D\chi \exp[\frac{1}{2}(\chi, A\chi) - \frac{1}{2}(\varphi, B\varphi) + (\chi, C\varphi) \\ - \frac{g}{4}\sum_k (\varphi_k, \Gamma\varphi_k)(\chi_k, \Gamma\chi_k)] \,. \qquad (A15)$$

Using an auxiliary spin field $\sigma_k = \pm 1$, the quartic terms in (A14), (A15) are decoupled

$$\exp[\frac{g}{4}(\varphi, \Gamma\varphi)(\psi, \Gamma\psi)] = \frac{1}{2}\sum_{\sigma=\pm 1} \exp\{\sigma\sqrt{g}/2[(\varphi, \Gamma\varphi) + (\psi, \Gamma\psi)]\} \,,$$

$$\exp[-\frac{g}{4}(\varphi, \Gamma\varphi)(\chi, \Gamma\chi)] = \frac{1}{2}\sum_{\sigma=\pm 1} \exp\{\sigma\sqrt{g}/2[-(\varphi, \Gamma\varphi) + (\chi, \Gamma\chi)]\}$$

$$(A16)$$

and the problem is reduced to the preceding one:

$$Z_a = Z_c = 2^{-N} \sum_{\{\sigma\}} \det{}^{1/2}(A + \sigma_k \sqrt{g}\Gamma) \int D\varphi \exp\{\frac{1}{2}(\varphi, B\varphi)$$
$$+ \frac{1}{2}(\varphi, C^T(A + \sigma_k \sqrt{g}\Gamma)^{-1}C\varphi)\} \qquad (A17)$$
$$= 2^{-N} \sum_{\{\sigma\}} \det{}^{1/2}\begin{pmatrix} A + \sigma_k \sqrt{g}\Gamma & C \\ -C^T & B + \sigma_k \sqrt{g}\Gamma \end{pmatrix}.$$

Note that the substitution (A16) is valid for two-component fields. When there is a great number of components an auxiliary summation should be introduced in a different manner: either the spin variable σ runs a greater number of values or we use several spin variables for every site. The result still holds.

References

1. Onsager L.: Phys.Rev. **65** 117–129 (1944)
2. Wannier G.H.: Phys.Rev. **79** 357–364 (1950)
3. Baxter R.J.: Phys. Rev. Lett., **26** 832–934 (1971)
4. Wu F.Y.: Phys.Rev. **B 4** 2312–2314 (1971)
5. Dalton N.W., Wood D.W.:J.Math. Phys. **10** 1271–1302 (1969)
6. Fun C., Wu F.Y.: Phys. Rev. **179** 560–570 (1969)
7. Gibberd R.V.:J. Math. Phys. **10** 1026–1029 (1969)
8. Binder K., Landau D.P.: Phys. Rev. **B 21** 1941–1962 (1980)
9. Bugrij A.I., Schadura V.N.: Physics of Many-Particle Systems, **12** 85–95 (1987)
10. Schultz T.D., Mattis D.S., Lied E.H.: Rev. Mod. Phys., **36** 856–867 (1964)
11. Fradkin E.S., Steingradt D.M.: Nuovo Cim. **47 A** 115–138 (1978)
12. Bugrij A.I.: Physics of Many-Particle Systems **13** 72–80 (1988)
13. Polubarinov I.V.:Preprint JINR, E17-413, 1984, Dubna.
14. Plechko V.N.:Dokl. Akad. Nauk SSSR **281** 834–837 (1985)
15. Bugrij A.I.:Preprint ITF-85-114R, 1985, Kiev.
16. Berezin F.A.: Usp. Mat. Nauk **24** 3–22 (1969)
17. Fradkin E.S. in *Problems of Theoretical Physics* (Nauka, Moscow 1969) pp.386–392
18. Plechko V.N.:Teor. Mat. Fiz. **64** 150–162 (1985)
19. Vaks V.G., Larkin A.I., Ovchinnikov Yu.N.: J. Teor. Eksper. Fiz. **49** 1180–1189 (1965)
20. Landau D.P.:J. Appl. Phys. **42** 1284–1285 (1971)
21. Fan C., Wu F.Y.:Phys. Rev. **B 2** 723–733 (1970)
22. Hantmacher F.R.: *Theory of Matrices* (Nauka, Moscow 1966) p.59

Ferromagnetism of Charge–Transfer Crystals: Curie Temperature of a Organometallic Ferromagnet

A.L. Tchougreeff and I.A. Misurkin

Karpov Institute of Physical Chemistry, 103 064 Moscow K-64, USSR

1. Introduction

Recently J.S. Miller and his coworkers have reported on the discovery of the ferromagnetic order in the organometallic charge–transfer salt of decamethylferrocenium (DMFc) and tetracyanethylene (TCNE). The ordered state occurs at temperatures lower than $T_c = 4.8\,\text{K}$ [1-6] and it is characterized by extremely large internal magnetic field (up to 450 kG) measured in the ^{57}Fe Mössbauer experiments.

In the present paper we use the model of ordered states of quasi-one-dimensional crystals [7] to build up a theory of the ferromagnetic order in DMFc–TCNE.

2. Theory

We begin from the description of electronic structure of the DMFc and TCNE molecules. TCNE is a molecule with closed electronic shell. It is also a good electron acceptor. The DMFc molecule is a d^6-complex. So the six of its electrons are placed on the strongly localized nonbonding $3d$-orbitals of the Fe atom. The other electrons occupy the delocalized bonding orbitals of the $C_5(CH_3)_5$ ligands and $4s$ and $4p$ orbitals of Fe. It follows from the analysis [8] of the spectroscopic data that the nondegenerate $3d_{z^2}$-orbital of Fe is the highest occupied molecular orbital of DMFc.

In the DMFc–TCNE crystal the DMFc and TCNE molecules alternate and form stacks:

$$rmDMFc^+$$
$$rmTCNE^-$$
$$rmDMFc^+$$
$$rmTCNE^-$$

which are oriented in the a-direction. The nonbonding $3d_{z^2}$-orbital of each DMFc molecule is occupied by single unpaired electron. In the same manner the LUMO of each TCNE is occupied by an electron. It means that the radical–ions TCNE$^-$ and the d^5-complex DMFc$^+$ both having total spin 1/2 present in the stack.

In DMFc there are two systems of orbitals. The first one consists of the localized d-orbitals of Fe^{3+} ions. The second one is composed of the delocalized π-orbitals of two C$_5$(CH$_3$)$_5$ ligands and the overlapping $4s$ and $4p$ orbitals of Fe. The second system of orbitals will be referred to as the *organic orbitals* of DMFc. To describe d-electrons (d-*system*) it is sufficient to take into account one $3d_{z^2}$-orbital per Fe atom. To describe delocalized electrons (*organic system*, OS) in the stack we take into account the orbital occupied by the unpaired electron on each TCNE$^-$ and the lowest unoccupied *organic orbital* (LUOO) of each DMFc$^+$.

The model Hamiltonian of the DMFc–TCNE stack is

$$H = -\sum_{n,\sigma}(\alpha_1 c^+_{1n\sigma}c_{1n\sigma} + \alpha_2 c^+_{2n\sigma}c_{2n\sigma})$$
$$- t_{\|}\sum_{n,\sigma}(c^+_{1n\sigma}c_{2n\sigma} + c^+_{2n\sigma}c_{1n+1\sigma} + \text{h.c.})$$
$$+ \gamma_1 \sum_n c^+_{1n\alpha}c_{1n\alpha}c^+_{1n\beta}c_{1n\beta} + \gamma_2 \sum_n c^+_{2n\alpha}c_{2n\alpha}c^+_{2n\beta}c_{2n\beta} \quad (1)$$
$$- W \sum_{n,\tau} A^+_{n\tau}A_{n\tau} + K \sum_n \sum_{\sigma\tau}\sigma\tau c^+_{1n\sigma}c_{1n\sigma}A^+_{n\tau}A_{n\tau}.$$

Here $c^+_{1n\sigma}$ is the creation operator of an electron on the LUOO of DMFc, $c^+_{2n\sigma}$ is the creation operator of an electron on TCNE, $A^+_{n\tau}$ is the creation operator of an electron on the $3d_{z^2}$ DMFc orbital; n is the DMFc–TCNE unit number in the stack. The electron spin projections $\sigma, \tau = \pm 1/2$ are denoted by the subscripts α and β respectively. The Hamiltonian (1) includes the attraction energy of an electron in the OS to the cores of DMFc$^+$ and TCNE, which are proportional to α_1 and α_2 respectively, the electron hopping in the OS, the Coulomb repulsion of the electrons with opposite spin projections occupying the same orbital, the attraction of d-electrons to core and the exchange interaction of d-electrons with the electrons in the OS inside a DMFc$^+$ cation.

Note that the stacks are not infinite. Structural defects break them into fragments of finite length. Therefore we consider finite fragment but use the cyclic boundary conditions. The ground state wavefunction can be represented by an antisymmetrized product of the wavefunction of localized electrons in the d-system Φ_1 and of the wavefunction of delocalized electrons in the OS Φ_2:

$$\Phi = \Phi_1 \wedge \Phi_2$$
$$\Phi_1 = \prod_n A^+_{n,-S/2}|0\rangle \;, \quad \Phi_2 = \prod_{|k|<k_F} g^+_{k\beta}g^+_{k\alpha}|0\rangle \;, k_F = \pi/2 \;. \quad (2)$$

In the UHF approximation for the OS we have the selfconsistent conditions for the operators $g^+_{k\sigma}$, which describe the SDW state in the OS. The OS spin density on DMFc is Q_{spin} and the spin density on TCNE is $-Q_{\text{spin}}$. The ground state of the stack has the total spin $N/4$, where N is the number of DMFC–TCNE units, because all the spins of d-electrons are oriented in the same direction due to interaction with spin density in the OS. Because the total spin of the stack turns out to be proportional to its length we can say that its ground state is ferromagnetically ordered.

The spin density Q_{spin} on the LUOO of DMFc turned out to be about 0.10. This spin density gives the Fermi contact contribution (H_{FC}) to the internal magnetic field on the ^{57}Fe nucleus. It is as follows:

$$H_{FC} = \frac{8\pi g \mu_B}{3} C^2_{4s} |\Psi_{4s}(0)|^2 Q_{\text{spin}} ,$$

where g is the electronic g-factor, μ_B is the Bohr magneton, C^2_{4s} is the weight of $4s$ atomic orbital in the LUOO of DMFc (it equals to 0.75 [10]), $|\Psi_{4s}(0)|^2$ is the density of $4s$ atomic orbital on the nucleus (it equals to $3.042 a_0^{-3}$ (a_0 is the Bohr radius) [9]). Substituting these values and the value of spin density in the expression for H_{FC} one obtains that the Fermi contact contribution is 330 kG. The Fermi contact contribution of $4s$ electrons has the same sign as the Fermi contact contribution of the core ($1s$, $2s$, and $3s$) electrons of the Fe atom. For the compound with single unpaired d-electron per Fe atom the latter is about 110 kG [6, 11], and the total internal field becomes 440 kG which is in agreement with the experimental value (~ 450 kG) [2, 6].

The spin density alternates in the OS. Let S be the phase of the SDW in the OS. It can be either +1 or -1, which corresponds to the different signs of the spin density on an orbital in the OS and to the different projections of the total spin of the stack.

Consider the ground state of a system composed of two stacks. In a line with [7] we describe each fragment by the Hamiltonian (1) and set S_i to be the SDW phase of the fragment i. Let the fragments be of equal length and contain N DMFc–TCNE units. Consider two adjacent fragments i and j of length N shifted by M DMFc–TCNE units, as on the scheme:

$$|\longleftarrow \quad N \quad \longrightarrow|$$

```
    ┌─────────────────┐
    │        i        │
    └─────────────────┘
         ┌─────────────────┐
         │        j        │
         └─────────────────┘
```

$$\longrightarrow| \quad M \quad |\longleftarrow$$

The resonance interaction operator for the pair is:

$$H_I = -t_\perp \sum_{n,\sigma} [c^+(i)_{1n\sigma} c(j)_{2n+M\sigma} + c^+(i)_{2n\sigma} c(j)_{1n+M\sigma}] + \text{h.c.} , \qquad (3)$$

where $t_\perp$ is the parameter of interstack electron hopping between the LUMO of TCNE and the LUOO of DMFc.

Treating the resonance interaction (3) as a perturbation (as in [7]) we find the second-order correction to the ground state energy. The numerical estimation shows that the energy of the pair is minimal when $S_i = S_j$.

$$\Delta E^{(2)}(S_i, S_j) = -(\nu^2 t_\perp^2/W) S_i S_j = -J S_i S_j\,, \quad \nu = (N-M)/N\,,$$

where W is effective bandwidth of the OS. This means that the interfragment interaction introduced above leads to the ferromagnetic state for the pair of fragments.

Consider now the quasi-one-dimensional crystal, and take into account only the resonance interaction between fragments. In the ferromagnetic crystals [2, 5, 6] the stacks are oriented along the a axis. According to [6], the mutual arrangement of adjacent stacks in the DMFc-TCNE crystals may be either in-registry or out-of-registry by one-half the chain axis length:

	$\leftarrow$ 8.7 Å $\rightarrow$			$\leftarrow$ 8.0 Å $\rightarrow$
	DMFc$^+$	DMFc$^+$		DMFc$^+$	TCNE$^-$
	TCNE$^-$	TCNE$^-$		TCNE$^-$	DMFc$^+$
	DMFc$^+$	DMFc$^+$		DMFc$^+$	TCNE$^-$
$a\ \uparrow$	TCNE$^-$	TCNE$^-$	$a\ \uparrow$	TCNE$^-$	DMFc$^+$
	in-registry			*out-of-registry*

The interstack electron hopping for the *in-registry* arrangement couples molecules of the same sort. In the case of *out-of-registry* arrangement the molecules of different sorts are coupled. The distance between axes of the stacks in the latter case is shorter than in the former one. Taking into account the exponential decrease of the hopping parameter, we neglect the interstack electron hopping between the *in-registry* arranged stacks. The interstack interaction for the case of *out-of-registry* arrangement is described by the resonance operator (3). According to the structural data on the Miller ferromagnet [4, 6], there are only four adjacent *out-of-registry* stacks for every given DMFc-TCNE stack. To estimate the critical temperature T_c, we assume (as in [7]) that all the fragments are of the same length N, and that the defects are displaced regularly so that the fragments of adjacent stacks are shifted relative to one another by $M = N/2$ DMFc-TCNE units. Now each fragment has eight equivalent neighbors, and the entire crystal may be presented as a body centered cubic (bcc) lattice with fragments of length N in sites.

Treating the interfragment resonance interaction as a perturbation, we obtain the second-order correction to the energy of the entire crystal. It is the sum of the contributions from all the pairs of interacting fragments (6):

$$\Delta E^{(2)} = \text{const} - J \sum_{m,l} S_m S_l\,. \qquad (4)$$

The SDW phases S_m can be regarded as variables of the Hamiltonian of the Ising model with ferromagnetic interaction between nearest neighbors. In the ground state of such a system the values of all the spin variables S_m are equal. This corresponds to the ferromagnetic state of the entire crystal because the magnetic moment turns out to be proportional to the volume of the crystal. This state is broken down by the thermal fluctuations if $T > T_c$.

To estimate the Curie temperature T_c we assume (in line with [7]) it to be equal to the critical temperature of the Ising model. For the latter, on bcc lattice there is a well-known relation [12]:

$$T_c = 6.32J \ . \tag{5}$$

Setting $t_\perp = 50\,\text{K}$, $N = 30$, and carrying out calculations with use of (1–5), one obtains $T_c = 5.4\,\text{K}$ in agreement with experiment [6]. The estimate $N = 30$ is given as for the Bechgaard salt [7]. Because the defects are randomly distributed, the regular Ising model is only first approximation to the problem.

3. Discussion

The results obtained in the present work make it possible to explain some phenomena observed in the crystals of charge-transfer complexes of metallocenes with organic acceptors.

The charge-transfer salt of DMFc with hexacyanobutadiene $[C_4(CN)_6]$ (HCNB) is a perfect electronic and structural analog of DMFc–TCNE. According to [5, 6], the magnetic susceptibility of DMFc–HCNB exhibits ferromagnet-type temperature dependence. Nevertheless the bulk ferromagnetism (spontaneous magnetization) does not take place in this system [1, 5, 6]. From our point of view, this can be explained by the value of T_c being extremely low. The electron affinity of HCNB is greater than that of TCNE, and the same applies to the absolute value of $\Delta\alpha$. Analysis of our model shows that the greater is $|\Delta\alpha|$, the smaller are the effective exchange integral J (5) and the critical temperature T_c.

4. Conclusion

In the present paper it is shown that it is important to take into account the fragmental structure of the stacks and the resonance interaction between the fragments in order to explain the ordering in the charge-transfer ferromagnets. This viewpoint differs significantly from the heuristic model [1–6] of the mechanism of the ordering in Miller ferromagnets.

Acknowledgements. The authors are grateful to Drs. V.Ya. Krivnov, A.M. Berezhkovskii, and A.V. Soudackov for invaluable discussions, and to Dr. A.Yu. Cohn for help during the preparation of the manuscript.

References

1. J.S. Miller, P.J. Krusik, D.A. Dixon, W.M. Reiff, J.H. Zhang, E.C. Anderson, A.J. Epstein: J. Am. Chem. Soc. **108** 4459 (1986)
2. J.S. Miller, J.C. Calabrese, H. Rommelmann, S.R. Chittipeddi, J.H. Zhang, W.M. Reiff, A.J. Epstein: J. Am. Chem. Soc. **109** 769 (1987)
3. D.A. Dixon, J.S.Miller: J. Am. Chem. Soc. **109** 3656 (1987)
4. J.S. Miller, A.J. Epstein: J. Am. Chem. Soc. **109** 3850 (1987)
5. J.S. Miller, J.H. Zhang, W.M. Reiff: J. Am. Chem. Soc. **109** 4584 (1987)
6. J.S. Miller, A.J. Epstein, W.M. Reiff: Chem. Rev. **88** 201 (1988)
7. A.L. Tchougreeff, I.A. Misurkin: Fiz. Tverd. Tela **30** 1043 (1988) [in Russian]
8. A.B. Lever: *Inorganic Electronic Spectroscopy* (Elsevier, Amsterdam 1984)
9. J. Danon in *Chemical Applications of Mossbauer Spectroscopy* V.I. Goldanskii, R.H. Herber (Eds.), (Academic Press, N.Y. 1968)
10. E.M. Shustorovich, M.E. Dyatkina: Zh. Strukt. Khimii **1** 109 (1960) [in Russian]
11. V.I. Goldanskii, E.F. Makarov in *Chemical Applications of Mossbauer Spectroscopy* V.I. Goldanskii, R.H. Herber (Eds.), (Academic Press, N.Y. 1968)
12. M.E. Fisher: J. Math. Phys. **4** 278 (1963)

Subject Index

Reference is made to the *first* pages of the relevant articles.

1-d Hubbard model 12

2-d dimer models 114
2-d Ising model 132

3-d dimer models 114
3-d membrane with impurities 121

$(A\text{-}B)_x$–polymers 73
anomalous transport 121
antiferromagnetic phase energy 32, 34

chain-length-dependence 62
charge-transfer crystals 152
CI 93
CI in π-systems 93
$CoCl_4^{2-}$ complexes 108
$CoCl_6^{4-}$ complexes 108
combined Peierls dielectrics 73
continuum Hamiltonian 73
correlation effects 129
 – energy 93
 – pairing 32
CuO plane of HTSO 23
Curie temperature 152
 – of organometallic ferromagnet 152
current carriers 41

d-d–spectra 108
diffusive controlled processes (DCP) 129
dimerization amplitude 67, 68
dimer models, 2-d 114
 –, 3-d 114
dynamical correlation 101

effective Hamiltonian method 106
electron correlation 93
electronic pairings 23
 – structure 106
Emery model 86
equipotential surfaces 37

experimental evidence for kinks 41
extension of Geminals problem 93
external magnetic field 48

FCI 93
FCI and spin flip operators 93
fermion charge fractionization 48
fermionization 132
ferromagnetism 152
 – of charge-transfer crystals 152
finite polymethine chains 101
forward scattering amplitude 152
full CI in π-systems 93
functional integral 135

Geminals problem 93
generalized 2-d Ising model 132
Grassmann field 133
 – variables 135
ground state energy 114

half-filled conduction band 32
high-T_c superconductor oxides 41
HMO 101
Hubbard model 12, 86, 114
 1-d – 12
 low energy physics of – 12
 magnetic properties of – 114
 – with infinite interactions 114
Huckel molecular orbital (HMO) 101

induced charge 48
infinite interactions 114
instabilities 62
Ising model 132, 135
 2-d – 132

Keldysh diagram technique 73
kink 41
 – photo-generation 41
 – properties in high-T_c oxides 41
 experimental evidence for – 41

La-Sr-Cu-O metal oxide 32
Landau Luttinger liquid 12

magnetic properties 114
– of Hubbard model 114
many-body reactive systems 129
many-electron systems 86
mean-field study 23
mean kinetic energy 62
membrane with impurities, 1-d 121
method of cyclic permutations 86
Mott and Peierls instabilities 62
Mott-Peierls semiconductor 62

nearest-neighbor interactions 135
next-nearest neighbor interactions 135
noncompact surface 48
nonlinear optical properties 73
– susceptibility 73, 80
nonlinear susceptibility 73,
n–site model 114
n–site segments 114
numerical calculations 62

optical excitation energies 41
optical spectra 106
organic conductors 62
organometallic ferromagnet 123

pair interactions 135
particle spectra 12
partition function 135
Peierls (and Mott) instabilities 62
Peierls deformation 62

Peierls dielectrics 73
phase diagram 114
phase energy 32
PIA spectra 41
polarization scheme 93
polaron states 86
– in the Emery model 86
polymethine dyes (PMD) 101

representation parameters 135

singlet pairing energy 32
singlet state 93
small π-systems 93
spin flip operators 93

thin disordered layers 121
third harmonic generation 73
transition metal complexes (TMC) 106
   optical spectra of – 106
   electronic structure of – 106
transition metals 106
transport through thin disordered layers 121
triplet state 93
twin structure 54
– in the 1-2-3 system 54
two-chain model 114

uniaxial pressure 54

variational operator approach 93

Y-Ba-Cu-O metal oxide 32

Index of Contributors

Bandrauk, A.D. 80
Bugrij, A.I. 135
Burlatsky, S.F. 121, 129

Carmelo, J. 12
Cheranovskii, V.O. 86, 114
Chernoutsan, A.I. 121

Dakhnovskii, Yu.I. 73, 80
Dyadusha, G.G. 100

Ivanov, V.V. 93

Krivnov, V.Ya. 86, 114

Luzanov, A.V. 93

Misurkin, I.A. 106, 152

Nikolayev, V.S. 54

Oshanin, G.S. 114, 121, 129

Ovchinnikov, A.A. 1, 12, 23, 86, 114, 129
Ovchinnikova, M.Ya. 23

Peash, Yu.F. 93
Pokhodnia, K.I. 41
Ponezha, E.A. 32
Pronin, K.A. 73

Repyakh, I.V. 100

Sheinkman, M.K. 41
Shramko, O.V. 62
Sitenko, Yu.A. 48
Soudackov, A.V. 106

Tchougreeff, A.L. 106, 152

Ukrainskii, I.I. 1, 32, 41, 62

V. E. Zakharov (Ed.)

What is Integrability?

With the contributions by numerous experts

1991. XIV, 321 pp. 1 fig. (Springer Series in Nonlinear Dynamics) Hardcover
ISBN 3-540-51964-5

This monograph deals with integrable dynamic systems with an infinite number of degrees of freedom. Leading scientists were invited to discuss the notion of integrability with two main points in mind:
1. a presentation of the various recently elaborated methods for determining whether a given system is integrable or not;
2. to understand the increasingly more important role of integrable systems in modern applied mathematics and theoretical physics.
Topics dealt with include: the applicability and integrability of "universal" nonlinear wave models (Calogero); perturbation theory for translational invariant nonlinear Hamiltonian systems (in 2+1d) with an additional integral of motion (Zakharov, Schulman); the role of the Painlevé test for ordinary (Ercolani, Siggia) and partial differential (Newell, Tabor) equations; the theory of integrable maps in a plane (Veselov); and the theory of the KdV equation with non-vanishing boundary conditions at infinity (Marchenko).

A. Hasegawa

Optical Solitons in Fibers

2nd enl. ed. 1990. XII, 79 pp. 25 figs. Softcover ISBN 3-540-51747-2

Already after six months high demand made a new edition of this textbook necessary. The most recent developments associated with two topical and very important theoretical and practical subjects are combined: **Solitons** as analytical solutions of nonlinear partial differential equations and as lossless signals in dielectric **fibers.** The practical implications point towards technological advances allowing for an economic and undistorted propagation of signals revolutionizing telecommunications. Starting from an elementary level readily accessible to undergraduates, this pioneer in the field provides a clear and up-to-date exposition of the prominent aspects of the theoretical background and most recent experimental results in this new and rapidly evolving branch of science. This well-written book makes not just easy reading for the researcher but also for the interested physicist, mathematician, and engineer. It is well suited for undergraduate or graduate lecture courses.

B. N. Zakhariev, A. A. Suzko

Direct and Inverse Problems
Potentials in Quantum Scattering

1990. XIII, 223 pp. 42 figs. Softcover ISBN 3-540-52484-3

This textbook can almost be viewed as a "how-to" manual for solving quantum inverse problems, that is, for deriving the potential from spectra and/or scattering data. The formal exposition of inverse methods is paralleled by a discussion of the direct problem.

In part differential and finite-difference equations are presented side by side. A variety of solution methods is presented. Their common features and (dis)advantages are analyzed. To foster a better understanding, the physical meaning of the mathematical quantities are discussed in detail.

Wave confinement in continuum bound states, resonance and collective tunneling, and the spectral and phase equivalence of various interactions are some of the physical problems covered.

V. B. Matveev, M. A. Salle

Darboux Transformations and Solitons

1991. IX, 120 pp. 12 figs. (Springer Series in Nonlinear Dynamics)
Hardcover ISBN 3-540-50660-8

In 1882 Darboux proposed a systematic algebraic approach to the solution of the linear Sturm-Liouville problem. In this book, the authors develop Darboux's idea to solve linear and nonlinear partial differential equations arising in soliton theory: the nonstationary linear Schrödinger equation, Korteweg-de Vries and Kadomtsev-Petviashvili equations, the Davey-Stewartson system, Sine-Gordon and nonlinear Schrödinger equation, 1+1 and 2+1 Toda lattice equations, and many others.

By using the Darboux transformation one can construct and examine the asymptotic behavior of multisoliton solutions interacting with an arbitrary background. In particular, the approach is useful in systems where an analysis based on the inverse scattering transform is more difficult.

The approach involves rather elementary tools of analysis and linear algebra so that it will be useful not only for experimentalists and specialists in soliton theory, but also for beginners with a grasp of these subjects.

Springer-Verlag
Berlin
Heidelberg
New York
London
Paris
Tokyo
Hong Kong
Barcelona
Budapest

Research Reports in Physics

The categories of camera-ready manuscripts (e.g., written in T_EX; preferably both hard and soft copy) considered for publication in the **Research Reports** include:

1. Reports of meetings of particular interest that are devoted to a single topic (provided that the camera-ready manuscript is received within four weeks of the meeting's close!).
2. Preliminary drafts of original papers and monographs.
3. Seminar notes on topics of current interest.
4. Reviews of new fields.

Should a manuscript appear better suited to another series, consent will be sought from the author for its transfer to the other series.

Research Reports in Physics are divided into numerous subseries, e.g., nonlinear dynamics or nuclear and particle physics. Besides covering material of general interest, the series provides the possibility for topics that are too specialized or controversial to be published within the traditional avenues. The small print runs make a consistent price structure impossible and will sometimes have to presuppose a financial contribution from the author (or a sponsor). In particular, in the case of proceedings the organizers are expected to place a bulk order and/or provide some funding.

Within **Research Reports** the timeliness of a manuscript is more important than its form, which may be unfinished or tentative. Thus in some instances, proofs may be merely outlined and results presented that will be published in full elsewhere later. Since the manuscripts are directly reproduced, the responsibility for form and content is mainly the author's.

Springer-Verlag
Berlin
Heidelberg
New York
London
Paris
Tokyo
Hong Kong
Barcelona
Budapest

☐ Heidelberger Platz 3, W-1000 Berlin 33, F. R. Germany ☐ 175 Fifth Ave., New York, NY 10010, USA ☐ 8 Alexandra Rd., London SW19 7JZ, England ☐ 26, rue des Carmes, F-75005 Paris, France ☐ 37-3, Hongo 3-chome, Bunkyo-ku, Tokyo 113, Japan ☐ Room 701, Mirror Tower, 61 Mody Road, Tsimshatsui, Kowloon, Hong Kong ☐ Avinguda Diagonal, 468-4° C, E-08004 Barcelona, Spain

Research Reports in Physics

Manuscripts should be no less than 100 and no more than 400 pages in length. They are reproduced by a photographic process and must therefore be typed with extreme care. Corrections to the typescript should be made by pasting in the new text or painting out errors with white correction fluid. The typescript is reduced slightly in size during reproduction; the text on every page has to be kept within a frame of 16 × 25.4 cm (6⁵⁄₁₆ × 10 inches). On request, the publisher will supply special stationery with the typing area outlined.

Editors or authors (of complete volumes) receive 5 complimentary copies and are free to use parts of the material in later publications.

All manuscripts, including proceedings, must contain a subject index. In the case of multi-author books and proceedings an index of contributors is also required. Proceedings should also contain a list of participants, with complete addresses.

Our leaflet, *Instructions for the Preparation of Camera-Ready Manuscripts,* and further details are available on request.

Manuscripts (in English) or inquiries should be directed to

Dr. Ernst F. Hefter
Physics Editorial 4
Springer-Verlag, Tiergartenstrasse 17
W-6900 Heidelberg, Fed. Rep. of Germany
(Tel. [0] 62 21-48 74 95;
Telex 461 723;
Telefax [0] 62 21-41 39 82)

☐ Heidelberger Platz 3, W-1000 Berlin 33, F. R. Germany ☐ 175 Fifth Ave., New York, NY 10010, USA ☐ 8 Alexandra Rd., London SW19 7JZ, England ☐ 26, rue des Carmes, F-75005 Paris, France ☐ 37-3, Hongo 3-chome, Bunkyo-ku, Tokyo 113, Japan ☐ Room 701, Mirror Tower, 61 Mody Road, Tsimshatsui, Kowloon, Hong Kong ☐ Avinguda Diagonal, 468-4° C, E-08006 Barcelona, Spain